Tomatoes & Basil on the 5th Floor

All you need for a bountiful balcony harvest

Patrick Vernuccio

CONTENTS

FOREWORD

This book stems from hard work and passion. In each paragraph, I've tried to put myself in your shoes, asking: "Is this easy to follow, useful, and doable?" When you're just starting out, you often read everything and anything about a subject, which can be confusing and even discouraging. So I've done my best to share my experience and to make some of the more complex concepts both practical and easy to remember.

Like a garden plant, this book took on a life of its own, evolving with the turbulent ups and downs of national and international news: the Intergovernmental Panel on Climate Change 2022 report warning of "unequivocal evidence" that humans are responsible for climate change (see page 141); intense heatwaves followed by flooding; and the tragic war in Ukraine with its threat to global food security. Even with all this going on, now is not the time to be fatalistic – we must act and create a virtuous circle full of hope.

These events led me to shift my approach to this book towards more sustainable methods, such as reusing water, recycling materials for the vegetable garden, reusing potting compost, and making your own fertilizers, compost, and so on.

Even if you live in an urban area, there are many resources you can reuse to enrich your soil once you take the time to look around and find out about them.

In the age of the fabled metaverse, it's more important than ever to open our eyes to our beautiful planet, to better appreciate, protect, and understand it. Starting a vegetable garden is part of advocating for nature, your loved ones, and future generations. Rather than calling these generations X, Y, or Z, I sincerely hope that the next one can be called a "Re-generation".

I hope you enjoy this book – may your reading be full of dreams, wonder, and many moments of connection in your vegetable garden!

Green love!
Patrick

GROW, LITTLE BASIL

So you want to start a vegetable garden on your balcony or terrace? Congratulations on this brilliant idea – it might be a gamechanger in your everyday life!

You may be feeling doubtful: is it really possible to grow a vegetable garden on your balcony? Do you even have green fingers?

This book is here to show you how to get started – it's rooting for you (note: this is not the only gardening pun in this book...)! It's also designed to guide you through the seasons as you take your first steps as a gardener with a simple method for growing many different kinds of fruit and vegetables in pots and window boxes, all year round!

BASIL TRAUMA

You'll no doubt remember that lovely potted organic basil you bought at your local supermarket, which gave up the ghost after a few days. Yet it was planted in soil and you watered it frequently and diligently. For the more adventurous among you, you might even have decided to repot it in one of your own containers, hoping to give it a second chance. But to no avail, the basil just wouldn't revive... Was this its destiny, or even a curse? Of course not! You've simply suffered from "basil trauma", which left you thinking you don't have the knack of keeping plants alive, and probably never will. Take a deep breath – we're going to find a cure.

First, basil is an annual plant. This means it completes its life cycle over one year. That's its natural cycle, and it's quite short. Second, supermarket basil bought in pots is not meant to continue growing and developing; rather, it is meant to be eaten shortly after purchase. Third,

yes, you did repot it in soil and water it, but it was most likely in depleted soil. In general, new potting compost offers only six weeks' worth of nutrients. After that, you need to feed it; otherwise nothing will grow properly. And I can confirm that, no, those late-night glasses of wine spilled in your window box or crushed cigarette ends certainly didn't help!

MY PERSONAL EXPERIENCE

Three years ago, I knew absolutely nothing about gardening. I'd never planted a seed in my life, and even a cactus trembled at the sight of me! When the warmer weather came, I decided to go to a garden centre and pick up some flowers to spruce up my balcony. When I arrived, I found myself in front of a huge rack overflowing with all kinds of seeds – tomatoes, radishes, lettuces, aubergines, courgettes, and so on. A dream come true!

I love aperitifs and dinner parties with friends, so I thought it would be nice to have lettuces and tomatoes to eat straight from my balcony. Without giving it a second thought, I made an impulse purchase – undoubtedly the best one I've ever made! I grabbed a few pots, some random packets of seeds, and some potting compost, and then I headed home with my little haul, which I rushed to unpack.

Then I paused for a long, awkward moment... Where to begin? What was I supposed to do? Would this tiny lettuce seed really grow into the glorious lettuce depicted on the packet? How many radish seeds was I meant to plant? Would one radish seed sprout into a bunch of radishes, or into a single one?

Then I asked myself some of the bigger questions: how was it that at age 36, I was so ignorant in this matter? I quickly blamed school and my education. There hadn't been a single lesson about how to grow your own food! Then I thought about my urban lifestyle. It's so quick and easy to find everything at the supermarket that we don't even take the time to ask ourselves where our fruit and vegetables come from, or how they grow. Personally, I didn't know that a tomato or aubergine developed from a flower. I used to think courgettes grew under the ground, and I would never have imagined that chicory needs to grow in the dark.

After doing a bit of research, I took the plunge. I had access to a lot of information – too much in fact. I read everything and anything, which was more than confusing. This book will answer your questions by giving you the right information in a fun, accessible way, focusing on growing in pots and balcony planters. By following a few basic principles and trusting your intuition, it's easy to grow your own vegetable garden. We can all grow healthy produce: it's just a matter of practice!

WELCOMING NATURE INTO OUR LIVES AND CITIES

CHANGING HOW WE CONSUME

The UN states that by 2050, 68 per cent of the world's population will be living in urban areas. You don't need to be a Nobel Prize winner to understand that the more urban areas expand, the more rural areas shrink. If this equation is simple, its solution is less so: more people, a growing demand for food, but less arable land...

RETHINK AND INNOVATE

Faced with this reality, we need to rethink our cities, our habits, patterns, and the way we consume: in short, we must reorganize our urban lifestyles. Fortunately, many innovative solutions are being implemented, and urban agriculture is a real opportunity and a wonderful alternative.

Obviously, your balcony alone won't solve the problem. Nor will you become entirely self-sufficient, although you may be able to be so for a few months every year and with certain varieties of garden plants, depending on your surface area. The important thing here is to think in terms of quality, not quantity.

Growing your own vegetable garden, even on a small scale, is the beginning of a journey that will change your life, helping you become more aware of consumption patterns. You'll experience the magic of planting a tiny seed and watching it grow into a plant that will bring you a fresh, new kind of happiness in the form of a wonderful fruit or vegetable, with an incomparable taste.

LIVING THE GREEN LIFE

Vegetable gardens also affect our wellbeing. Colours have a direct impact on our brain and our behaviour. While grey, for example, leads to indecision, green, on the other hand, is a soothing colour, conveying a sense of freshness and

harmony, as well as improving focus and concentration. It is also said to contribute to better cognitive development in children. So let's forget that concrete grey, and bring some soothing green into our lives and cities!

CONNECTING WITH FOOD AND THE SEASONS

Like the air we breathe, the food we eat has a major impact on our health. It is an essential part of our body's ecosystem.

EAT HEALTHY, EAT SEASONAL!

We're incredibly lucky to experience four distinct seasons, and the abundant variety of fruits and vegetables they bring all year round.

Seasonal fruits and vegetables are essential to the human body's needs. For example, to cope with the cold in winter, our bodies need a substantial supply of vitamin C and minerals. Many winter vegetables contain these nutrients, including spinach, kale, turnips, pak choi, broccoli, cauliflower, and sorrel. In the summer, our bodies need water to withstand the heat: it's the perfect season to have tomatoes, cucumbers, courgettes, strawberries, and radishes. The good news is that you can grow all these plants on your balcony!

IT'S JUST COMMON SENSE

Creating your own garden will open your eyes to the big difference between the fruits and vegetables you find in supermarkets and your own produce. You will definitely be changing your eating and grocery shopping habits. I guarantee that the taste of a homegrown tomato from your own balcony has absolutely nothing in common with that of a supermarket tomato—you'll turn your back on the bland flavour, the wateriness, the sometimes mealy texture, and the utter lack of nutrients!

In fact, numerous studies show a drastic drop in the nutrient content of our fruit and vegetables just over the last 50 years (see page 141). The main response from lobbies and manufacturers is that this fact has no scientific basis: they argue that 50 years ago, technology and measurements were far less precise than they are today...

However, since science can't (or won't) give us all the answers, I've decided to trust my intuition. And my intuition tells me that a hybrid tomato smothered with pesticides that travels hundreds of miles in the cold just can't compare in terms of nutrients with the tomato I've grown on my balcony where it got plenty of sun, water, and quality potting compost!

GO LOCAL

It is essential for us to return to local consumption patterns: we must create the regeneration our ecosystems so desperately need, just as we need to make our lifestyles and ways of thinking more sustainable.

THE REGENERATION PRINCIPLE

Regeneration is a fundamental concept of permaculture, which aims to implement a mode of agriculture based on sustainable principles that protect both the environment and people by following the patterns of natural ecosystems. The goal of permaculture is to regenerate nature, ecosystems, and all species, including humans.

Take a few minutes to think about what regenerates you in your life – the thing that, no matter how much effort and time you invest in it, keeps giving you a sense of happiness and fulfilment. It could be a hobby, an interest, a passion, a person, or anything in between. For me, unsurprisingly, it's the vegetable garden on my balcony! What about you?

TO THE FARMERS' MARKET!

Local farmers' markets play an active role in regeneration. A local market is a place where you see happy people, because it gives such a strong sense of community. In addition to making a purchase, you can talk to local farmers – you can experience a human connection. It's the kind of place that nurtures hope, where I see many people refusing paper or plastic bags, because you come to the market with your own shopping bag, sometimes your own glass jars.

Produce is also no longer displayed in plastic, but in its raw and natural form, right in front of us, allowing us to instantly reconnect with unfiltered seasonal fruits and vegetables.

While it may take some effort to shake up some of your urban habits, the reward for investing your time and money in a short, virtuous circuit and supporting local regeneration speaks for itself!

Your balcony is, of course, one of the shortest, most local food circuits there is. It will reconnect you with what's on your plate, and re-educate you about the seasons and the unique flavours of each fruit and vegetable.

URBAN AGRICULTURE

Urban and suburban agriculture means growing vegetables, fruit, and other foods in the city. This can be done on rooftops, balconies, and terraces, in courtyards, shared vegetable gardens, garages, and even public spaces.

Growing lettuce on your windowsill or microgreens indoors is a form of urban agriculture that we can all implement at our own scale.

TIME TRAVEL

In the 19th century, the city of Paris had a well-established system of urban agriculture. Back then, the city's market gardeners managed 1,378 hectares of arable land (3,405 acres), divided into 1,800 gardens. In 1843, two Parisian market gardeners, J.-G. Moreau and J.-J. Daverne, wrote the *Manuel pratique de la culture maraîchère de Paris* [*Practical Manual of Vegetable Cultivation in Paris*], published in 1845. It demonstrated the expertise people possessed to produce large quantities of quality vegetables in a small surface area, recovering waste and reusing it as fertilizer, thus reducing the risk of disease and the spread of pests. This book inspired Jean-Martin Fortier, a Quebecois farmer, teacher, and writer, to

develop a model for a small-scale vegetable farm or micro-farm. He details this model in his book, *The Market Gardener: A Successful Grower's Handbook for Small-Scale Organic Farming*, published in 2012.

Be inspired by history

Let's take a leaf out of the book of 19th-century market gardeners and repurpose urban waste to grow our own vegetables! Even on a balcony, it's totally possible to create a small compost heap, as well as reusing other household waste to create your own homemade fertilizers.

LET'S START A BALCONY VEGETABLE GARDEN

THE FRENCHIE GARDENER METHOD

Here we go! Let's get started with this balcony vegetable garden. It will change with the seasons as we try to bring in a variety of crops. With the right planning and nature's helping hand, you can even rotate the crops in your pots, so that your balcony stays green all year round!

THE TYPICAL BALCONY VEGETABLE GARDEN

My method is simple and easy to adapt for small spaces. The example I'll be using throughout this book is a 4sq m (43sq ft) balcony, half of which is exposed to direct sunlight all year round, and the other half to semi-shade. Of course, if you're lucky enough to have more space, you can add even more pots! There's plenty of time to stock up on pots and planters (see p. 48).

Containers

You'll need 7 pots, 5 troughs or window boxes, and 2 raised planters.

MY TIPS FOR MAKING YOUR LIFE EASIER

The "Saving time" boxes offer tips on where to find the supplies you need. The "Saving money" sections offer thrifty DIY alternatives with simple instructions.

WHAT SHALL WE GROW?

Spring and summer
- - - - - - - - - - - - - - - -

- **7 pots** cherry tomatoes, tomatoes, aubergines, peppers, cucumbers, potatoes, basil.
- **5 window boxes** strawberries, sorrel, phacelia, nasturtiums, zinnias.
- **2 raised planters** radishes, lettuce.

Autumn and winter
- - - - - - - - - - - - - - - -

- **7 pots** pak choi, kale, fennel, chard, parsley or coriander, cauliflower, broccoli.
- **5 window boxes** strawberries, sorrel, phacelia, lamb's lettuce, spinach.
- **2 raised planters** winter salad leaves and mizuna, turnips.

ADAPTING TO YOUR CLIMATE

Before running off to buy your planters, you should first research your local climate. This varies from one region to the next, and even from one city to the next, so much so that just a few miles into suburbia can produce a very different climate!

WHAT DO I NEED TO KNOW?

In gardening, you need to plan around two key dates: the last frost in spring and the first frost in winter. Depending on these dates, you'll need to plan everything and draw up your own calendar.

For example, some seed packets will tell you to plant certain garden plants in March. But if it's still freezing in March where you live, you'll have to wait a little longer, as some varieties die very easily if they're exposed to a light frost. So don't take any chances!

YOUR TURN!

Now that you know these two essential dates, you can plan your own sowing calendar starting a few weeks before the last spring frost date, and also know when to plant out your seedlings safely.

Where can you find out about your climate and frost dates?

Head over to plantmaps.com and click on UK maps for the last spring frost and first winter frost. Zoom in to find your precise location, or look for a town near you, and *voilà*, you now know when frost will begin and end! Remember to write down both of those frost dates.

City ....

Last spring frost ....

First winter frost ....

MANAGE YOUR SPACE

In the same way you need to know the size of the inside of your house or apartment, it's important to know the type of suitable space available on your balcony or terrace before you start setting up.

FINDING SUNLIGHT

One of the first things to do is determine your balcony's exposure to the sun. To do this, simply use the compass app on your smartphone, or better yet, a real compass! The free SunEarthTools website is also very useful. Simply enter your address, and you'll get the sun's position in the sky according to your space and the seasons. If you're facing a tall building, that doesn't mean you can't get any sun. It all depends on the sun's elevation!

LOOK AROUND YOU

Try looking around you too: a tree that's bare in winter will grow leaves in spring, perhaps providing you with a few hours of shade. So now you can try to map out the sunny and/or shady corners of your space through the year.

Just moved in?

In a new flat, a quick way to find out more about your outdoor space is to ask your neighbours to share their experiences. They'll be able to tell you about the exposure of their balcony, and quite possibly yours.

EASY WATERING

Another important factor to consider is your balcony's access to water. In more modern buildings, some balconies will have an outdoor tap that will make watering much easier, especially in the summer. Otherwise, you have two other options:

• **The good old 10-litre** (2.5 US gallon) watering can, to avoid trips to the sink.

• **If you have an indoor water tap close to your balcony**, all you need is a special nozzle to connect a garden hose. How convenient! Some nozzles even connect to the kitchen tap.

STAYING ON YOUR NEIGHBOURS' GOOD SIDE

When you water your plants, the water drains through the available drainage holes. If the water on your balcony drains into a gutter, then there is no problem. On the other hand, if it's likely to drip out into the street, I'm not sure your downstairs neighbours or passers-by will be happy about the surprise shower you're giving them! Plan ahead by always placing saucers under your flowerpots (see page 67).

Collecting rainwater

If your balcony has a gutter, you may be able to connect to it and collect rainwater in a barrel (see page 43). Just make sure you've got permission to do so, and that your balcony's foundations are strong enough to support the weight of a water tank.

USING VERTICAL SPACE

Now you know more about your climate, frost dates, and exposure, it's time to think about the layout of your space. The idea is to get to know it better, so you can build on its assets and find practical solutions for its drawbacks.

A MATTER OF SURFACE

The surface area of your balcony is determined by the floor space, measured in square metres or square feet. This is the surface you walk on, and on which you can place your pots, raised planters, and window boxes, as well as perhaps a small table and chairs. You may find that this area is too small to fit everything in.

REACH HIGHER

The key to urban gardening is to make the most of vertical space to accommodate more pots, but also to help your vegetable garden climb using trellises or stakes.

Look up and see how much vertical space you have. The height of walls and balcony railings, perhaps a fence separating you from your neighbour's balcony might be unused vertical space with the incredible potential for adding planters or pots, allowing you to grow even more plants!

Optimizing vertical space

In a small space, every bit can be put to good use, so why not furnish the walls? Before you start, check the structures to see if you can drill holes to fix a trellis or secure a shelf. If this isn't possible (building regulations may be restrictive), don't panic – there are other solutions. For example, you can measure your balcony railings and choose planters that can easily be attached to them, placing them so they face inwards.

AN IDEAL VEGETABLE GARDEN

Here's an example of a balcony vegetable garden that fully utilizes vertical space. Recycled and upcycled objects have been repurposed as pots. Whether it's crates, wooden pallets, water bottles, or water containers, you can use these objects to inspire you into giving them a new purpose and help you grow your very own jungle right outside the door to your balcony.

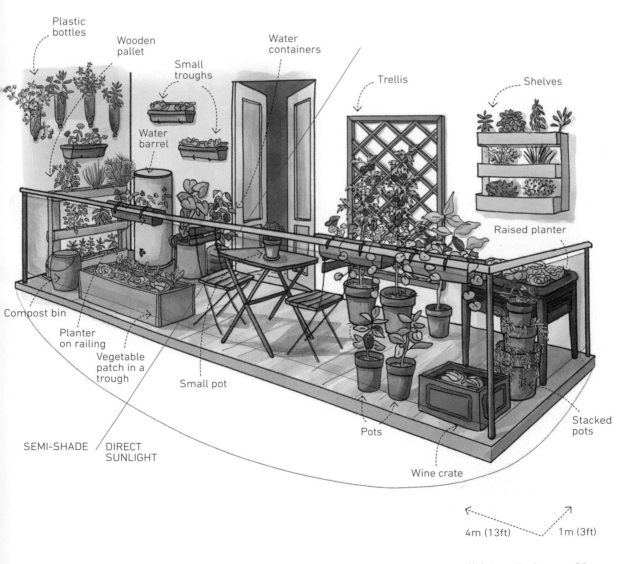

Plastic bottles

Wooden pallet

Small troughs

Water containers

Trellis

Shelves

Water barrel

Raised planter

Compost bin

Planter on railing

Vegetable patch in a trough

Small pot

Small pot

Pots

Stacked pots

SEMI-SHADE / DIRECT SUNLIGHT

Wine crate

4m (13ft)   1m (3ft)

SHORT GROWTH CYCLES

Most root vegetables (radishes, turnips) and leafy vegetables (lettuce, spinach) have a short growth cycle. From sowing to harvesting, you should allow between one and four months at most.

1. **Sowing** Start your seedlings indoors or plant seeds directly outdoors in pots and window boxes when the weather gets mild.

2. **Growth** Plant your seedlings outdoors. Your plants will then begin their growth phase, sprouting leaves and roots. You'll need to fertilize them to provide them with the necessary nutrients.

3. **Harvesting** Once the produce is ripe, you can harvest it!

4. **Flowering** If the vegetable hasn't been harvested, the plant will begin its reproductive phase and go to seed. The flowers will produce seeds that you can harvest and plant the following year. Nature is generous, and from a single lettuce plant, you'll be able to harvest a hundred seeds!

SHORT GROWTH CYCLE
1–4 months

SOWING - - -> GROWTH - - -> HARVESTING - - -> FLOWERING

LONG GROWTH CYCLES

Long-cycle plants include tomatoes, aubergines, peppers, cucumbers, and strawberries. From sowing to harvest, you'll need to allow around five to six months. These plants need time to grow strong before producing fruit.

1. **Sowing** Start your seedlings indoors in a warm room to encourage germination.

2. **Growth** Plant your seedlings outdoors once frost is no longer a threat. A light frost can be fatal, so it's best to be patient. The plants will then enter their true growth phase, sprouting leaves and roots. You'll need to feed them to provide the necessary nutrients.

3. **Flowering** Unlike the short cycle, flowering here does not mean the end of the plant's life. This stage is crucial, as the fruits or vegetables will grow from the flowers! You will again need to provide the right nutrients to encourage healthy flowers, and thus a good harvest.

4. **Harvesting** Once the flowers have been pollinated, they grow into fruits or vegetables. Then comes the happy harvesting phase! If you want to harvest your own seeds, don't look for them in the flowers, but in the fruit or vegetable. Simply cut open a tomato or aubergine to find them.

LONG GROWTH CYCLE
5–6 months

SOWING ---> GROWTH ---> FLOWERING ---> HARVESTING

LATE
WINTER

LATE WINTER ON YOUR BALCONY

Winter is a season some people look forward to and others dread. It's cold, and nighttime or even daytime frosts are commonplace. It's a good time to throw on your grandmother's thick sweaters and bundle up in hats and scarves, with a hot chocolate or mulled wine in hand!

HIBERNATION MODE

Even if it's not as cold as it is out in the countryside, winter in the city is quite different from other seasons. Days are shorter, there is less sunlight, everything sometimes seems to come to a standstill, and it can also feel that way in your balcony vegetable garden.

In the winter, everything grows more slowly, as plants focus most of their energy on survival. Surprisingly, some plants actually thrive, such as kale, spinach, and winter salads. Kale, in particular, is impressive for its resilience. It may freeze in sub-zero temperatures, but it will still stand proudly under the snow, bravely facing the cold. Frost can even affect it in a good way, softening the structure and texture of its leaves, which can be

SHOPPING AND REPURPOSING LIST

- Seeds
- Seedling soil
- Pots and cups for seedlings
- A small watering can and/or spray
- A propagator or seedling trays
- A bokashi compost bin
- Optional: LED seedling light
- Optional: water barrel

more palatable than in spring. So it's perfectly possible to have a balcony vegetable garden over winter, as long as you choose the right plants!

PLAY IT COOL

The other perk of winter gardening is that it's low maintenance. You don't need to water your plants as much, and pests are few and far between, so you can use the precious time you've saved to plan ahead for spring.

Late winter also means the start of a major phase: indoor sowing. You'll be able to watch your seeds starting to grow, sprouting seedlings from the soil. How wonderful!

LATE WINTER TO-DO LIST

- Continue to protect plants growing under domes, covers, and/or protective fleece
- Choose and order seeds
- Prepare sowing supplies
- Plan a sowing schedule
- Start sowing
- Care for seedlings
- Start composting
- Collect rainwater

CHOOSING THE RIGHT SEEDS

Winter hasn't quite ended yet, but spring is just around the corner! It's time to plan ahead for the season and put your best foot forward.

SEEDS EVERYWHERE!

Late winter is the ideal time for some daydreaming. It's still cold out, and you're probably cosied up under your favourite blanket... So why not let your mind wander and think about all the amazing plants you can grow when the warm weather comes back?

In the box on the right, I've shared some cultivars (named plants cultivated for certain characteristics) and plant suggestions that grow very well in pots, but feel free to try different cultivars and experiment. I highly recommend you visit your local garden centre, where you'll be able to browse many options! All the cultivars I suggest here are heirloom seeds rather than hybrids. I'll explain the main difference between the two next.

My favourite seeds

- **Aubergine** 'Listada de Gandia' or 'Little Fingers'
- **Basil** 'Eleonora' or 'Red Rubin'
- **Cherry tomato** 'Peacevine' or 'Verde Claro'
- **Cucumber** 'Le Généreux' or 'Mideast Peace'
- **Flowers** phacelia, nasturtiums, and zinnias
- **Lettuce** 'Reine de Mai' or 'Chêne frisée'
- **Pepper** 'Corne de bœuf' or 'Fushimi'
- **Potatoes** collect market potatoes to use as seed potatoes
- **Radish** 'Flamboyant' or 'Cherry Belle'
- **Tomato** 'Marmande' or 'Green Zebra'

HYBRID SEEDS VS. HEIRLOOM SEEDS

Vegetable seeds can be divided into two categories: the good, a.k.a. heirloom seeds; and the bad, a.k.a. hybrid seeds. There is a major difference between these seed types – a world of difference, in fact. Committing to grow heirloom seeds is the most sustainable choice to protect the environment.

WHAT ARE HYBRIDS AND HEIRLOOMS?

Hybrid seeds are produced by cross-pollinating different plants. They are generally tagged F1, or with a number 1 or 2 after the plant name on the seed packet. Heirloom seeds, by contrast, aren't often found in seed packets – these are seeds from old and historic cultivars that have been handed down through generations and preserved for their valuable qualities, such as flavour and hardiness.

WHAT HYBRIDS ARE HIDING

Let's take the example of a hybrid tomato. It will be the result of crosses between cultivar A, which may have the advantage of being more resistant to disease, cultivar B which will keep longer, and cultivar C which is famous for its high yield. It all looks very promising on paper, and you would expect the ultimate tomato. However, it's more like having a shiny new Ferrari ... with the engine of a Volkswagen Beetle!

Hybrid seeds can't really reproduce. In fact, even if they do produce viable seeds, these seeds will almost never yield the same cultivar as the initial combination. Instead, you'll get a random mix of cultivar A and C, or B and A, but never that ultimate tomato. Why? Simply because the hybrid seed process is a model created by and for the seed industry.

Heirloom seeds, by contrast, are naturally grown, and are sure to reproduce. So, whereas an heirloom seed will yield a hundred free seeds of the same cultivar, with a hybrid cultivar, your only choice will be to buy the seeds again if you want to get the exact same one.

NOT IN THE CATALOGUE

Seed companies have their own "official" seed catalogues. Interestingly, heirloom seeds are excluded from these, and in many countries there are bans or heavy restrictions on selling them. However, many local and national organizations fight to preserve, save, share, and sell heirloom seeds. These include Garden Organic in the UK, the Irish Seed Savers Association in Ireland, and the Organic Seed Alliance and Seed Savers Exchange in North America. Some organizations such as the Alliance of Native Seedkeepers (North America) are committed to preserving heirloom seeds as well as indigenous seed keeping practices and heritage.

PRESERVING BIODIVERSITY

There are over 15,000 tomato cultivars in the world, each with their own unique taste, colour, size, shape, and texture. By choosing only certain cultivars to produce hybrid seeds, we are deliberately erasing the incredible diversity of fruits and vegetables that nature offers us. Choosing heirloom seeds preserves our natural heritage, just as we must protect endangered animal species.

THE PERKS OF HEIRLOOM SEEDS

Nature does not put a price tag on its many gifts! By opting for heirloom seeds, not only will you have access to a wide range of cultivars, but you'll also reconnect with beautiful, nutritious, tasty, and authentic fruits and vegetables. Be gone, bland, mealy tomatoes!

Moreover, the heirloom seeds you will collect will have learnt from the parent plant. This genetic transmission means plants will be even more adapted to your local growing conditions.

I therefore strongly advise you to grow heirloom seed, as it's also important to support the seed-keeping organizations who are protecting our natural heritage. You can make a difference too!

Nutritional values

In just 60 years, tomatoes may have lost over 59 per cent of their vitamin C content, and a contemporary apple contains 100 times less vitamin C than it did 50 years ago. Back then, a single orange was enough to meet our daily vitamin A requirements; today, we need 21 to meet that same target! As for broccoli, it contains four times less calcium and six times less iron than it used to. Numerous studies, including those by Brian Halweil, show an alarming correlation between nutritional loss and the way our food is grown. We may be eating prettier food, but we're also eating significantly less nutritious and blander food.

WHY SOW YOUR OWN SEEDS?

By sowing your own seeds, you experience the entire garden plant journey from seed to seedling, from plant to harvest. When you sow your own seeds, you reap fruits and vegetables, and so much pride and joy!

WHY SOW INDOORS?

Most of the time, seeds are sown indoors (except in the middle of summer), so that they can grow in a protected space and be ready to be planted outdoors when the time is right. You will need to give them your full attention and provide a regular water supply and plenty of light.

BUDGET-FRIENDLY SEEDS

Sowing your own seeds is a thrifty way to grow fruits and vegetables, as seeds are much cheaper than garden-centre plants. Remember that a single plant grown from an heirloom seed you saved yourself the previous year can be worth the price of an entire packet of hybrid seeds.

SO MANY OPTIONS TO CHOOSE FROM

Aside from being the thriftier choice, sowing your own seeds allows you to choose from a wide range of different garden plants. If you buy your seedlings at a garden centre, you'll obviously have fewer to choose from, and they probably won't be from heirloom seeds. On the other hand, growing your own seedlings does require some time set aside for daily maintenance.

A time to sow, a time to grow

As a rule of thumb, a tomato or aubergine seedling will take around six to eight weeks to grow indoors, with the right maintenance. A lettuce seedling will be ready in three to four weeks and is much easier to care for.

PLANNING A SOWING SCHEDULE

It may come as a surprise to you, but a schedule is an absolutely necessary tool for gardening, just like your watering can or your rake!

THE LAST SPRING FROST DATE

A sowing calendar will vary from one region to the next, and even from one town to the next. By now, after reading "Adapting to your climate" (see page 17), you should have researched and written down the date of the frost where you live.

You will probably read instructions telling you to plant a particular cultivar in March or April – ignore them. Only by finding the last spring frost date will you be able to draw up your own sowing schedule, following your local climate for your balcony.

YOUR SOWING CALENDAR

Each sowing date is set by counting backwards from the last spring frost date.

* Strawberries take a long time to grow from seed. You can skip them for now: turn to page 91 to learn how to propagate your strawberry plants from a garden-centre seedling.

Your last spring frost date :		
Vegetables/fruits/flowers	Growth	Sowing date
Aubergine	8 weeks	
Basil	8 weeks	
Cucumber	4 weeks	
Lettuce	4 weeks	
Nasturtium	4 weeks	
Pepper	10 weeks	
Potato	6 weeks	
Sorrel	4 weeks	
Strawberry*	12 weeks	
Tomato and cherry tomato	6 weeks	
Radish, phacelia, zinnia	No need to sow: seeds will be planted directly in outdoor pots after the last frost date.	

PLANNING A SOWING CALENDAR

LAST SPRING
FROST DATE

12 WEEKS BEFORE	10 WEEKS BEFORE	8 WEEKS BEFORE	6 WEEKS BEFORE	4 WEEKS BEFORE
STRAWBERRY	PEPPER	AUBERGINE BASIL	POTATO TOMATO	CUCUMBER LETTUCE NASTURTIUM SORREL

SUCCESSFUL SOWING SUPPLIES

• **Propagator** This is a tray covered with a transparent lid to conserve moisture for your seedlings.
• **Seedling labels** To help you navigate your vegetable garden, buy wooden or plastic labels from garden centres.

MAKING OR REPURPOSING

• **Seeds** Many gardening enthusiasts exchange seeds or give them away for free. So don't hesitate to ask friends and family for seeds or use social media to get in touch with other gardeners or organizations. You can also collect seeds from the previous year's plants.
• **Seed planting compost** Although I recommend buying seed compost for successful growth, you can also make your own. All you have to do is take

BUYING YOUR OWN

• **Seeds** Check your local seed supplier for their selection of heirloom seeds. Take a look at their catalogue, and you'll be full of ideas and projects!
• **Seed planting compost** This is a soft, light potting soil that helps seeds to sprout properly. It can be found in garden centres.
• **Pots and containers** You'll need a plastic seed tray (preferably recycled), which you can reuse over several years. You can also opt for biodegradable fibre seed trays or pots.
• **Watering can** A small indoor watering can of about 2 litres (0.5 US gallons) will do the trick.
• **Spray** A houseplant spray is needed to gently moisten the soil.

the potting compost you've already used and sift it through a fairly coarse-meshed sieve. The result will be light, soft soil, free of roots and other debris. Add a handful of coffee grounds to provide nutrients, then mix.

• **Pots and dishes** You can repurpose a lot of household waste to make seedling pots: the centre of a toilet roll, yogurt pots, drinks cans, small flower pots, plastic bottles, various containers, egg cartons... Be creative and reuse what would have been thrown away. The only requirement is to drill three or four holes in the bottom of each container to allow water to drain out. Your pots must then be placed on a tray to catch excess water.

• **Watering can** You can use a plastic drinks bottle, with a capacity of 1 litre (about 34 fl oz) as a watering can. Poke small holes in the cap to recreate the effect of a watering can head.

• **Spray bottle** Use a window cleaner spray bottle, for example – make sure to clean it thoroughly beforehand!

• **Seed tray** Use a tray to hold your pots of seedlings. Make sure it is water-resistant.

• **Seedling labels** Reuse lolly sticks or cut strips out of discarded plastic containers.

GET YOUR SEEDLINGS STARTED

Now you're going to learn how to start your own seedlings step by step. Follow the instructions closely and you'll see just how easy it is!

READY, GET SET, PLANT!

1. Grab your seed tray. If you're reusing plastic cups, don't forget to poke holes in them to allow the water to drain.

2. Place it on a pot saucer or a plastic sheet, so that you don't get compost all over your floor.

3. Fill your seed tray to the top with seed planting compost.

4. Press the compost gently with your fingers to remove any gaps, then add more to the top.

5. Water the compost in each section. Once the water has drained away, the compost level will have dropped slightly and the compost will be more malleable.

6. Pack down the compost lightly.

7. Make a hole in the compost with a pencil or chopstick. If the seed is small (lettuce, tomato), the hole should be 1cm (½in) deep. If the seed is larger (courgette, aubergine, cucumber), the hole should be 2cm (¾in) deep.

8. Put a seed in the hole and make sure it falls inside. A dark seed can be hard to spot, so use a light to check it's at the base of the hole.

9. Using the pencil or chopstick, gently tap the seed to make sure it is in contact with the compost.

10. Gently close the hole with your fingers.

11. Water the soil again with a spray so it is moist.

12. Don't forget to add a label to remember which variety of seed you've just planted!

Place your seed tray in a room at around 20°C (68°F), near a window with plenty of light, or directly under an LED grow light placed 15–20cm (6–8in) above. Cover with a transparent lid to keep the temperature and humidity higher.

Now be patient! Avoid letting the compost dry out, otherwise the seeds won't sprout. Make sure to spray the soil every day (see page 40).

Germination times

- **Aubergine** 10 days to 2 weeks
- **Basil** 7 to 10 days
- **Cucumber** 7 to 10 days
- **Lettuce** 7 to 10 days
- **Nasturtium** 10 days
- **Peppers** about 2 weeks
- **Sorrel** 7 to 10 days
- **Strawberry** 4 to 6 weeks
- **Tomato and cherry tomato** 7 to 10 days

How many seedlings do I need?

Not all your seedlings will sprout – that's up to nature to decide, and it will depend on the seed. To make sure you grow enough seedlings, I recommend sowing extra seeds:

Aubergine You'll need 1 plant. Sow 3 seeds.

Basil You'll need 1 plant. Sow 3 seeds.

Cucumber You'll need 1 plant. Sow 3 seeds.

Lettuce You'll need 8 plants. Sow 16 seeds.

Nasturtium You'll need 3 plants. Sow 9 seeds.

Peppers You'll need 1 plant. Sow 3 seeds.

Sorrel You'll need 1 plant. Sow 3 seeds.

Strawberries You'll need 2 plants. Sow 4 seeds.

Tomato You'll need 2 plants. Sow 3 seeds per cultivar, a total of 6 seeds.

WHAT ABOUT SPUDS?

You won't be sowing any potato seeds to get your seedlings. Rather, if you expose a potato to sunlight, it will germinate and sprout "eyes", which look like little buds. Stems will then grow from these eyes, allowing the plant to produce other potatoes in the soil.

All you need to do is get two organic potatoes. Place them on an egg tray eyes upwards in a cool, shady spot. After a few weeks, small stems will appear, and you will soon be able to plant them after the last frost!

WHEN DO I SOW LETTUCE, SORREL, AND CUCUMBER?

As mentioned in the sowing schedule, these seeds are sown last. They need to be planted four weeks before the last frost date in your region. Depending on this date, you may sow in early spring rather than in late winter.

SEEDLING CARE

Do you see a little sprout? Congratulations! Your seedling has sprouted, and you've just grown a brand new plant. Now it's time to tend to your seedlings so they're ready for the outdoors.

DAILY SEEDLING CARE

To ensure that your seedlings grow well, there are two things to look out for every day:
• **Never let potting compost dry out** Always keep the soil slightly moist. You can keep the lid of your propagator closed to retain moisture and heat. If your pots are on a tray, pour water directly into the tray: the water will be absorbed up into your pots.
• **Keep them well lit** Turn on the LED grow light for the whole day at least, placing your seedlings about 15–20cm (6–8in) apart, or set them in the sunniest spot in your home.

My seedlings look funny...

Do your seedlings have long stems and fall over backwards? This is known as "etiolation". The cause is a lack of light: your seedlings are trying to stretch as far as possible to get enough light. Try moving them to a much brighter spot. You can still plant them later, you will just need to bury them deeper.

HELP, MY SEEDLINGS AREN'T GROWING!

• **Check their average germination time** Some seeds may take a few days longer, others a few days less (see page 36). If sprouting is over a week late, your seed probably won't sprout. That's up to nature. You still have time to start new ones!
• **Check your room temperature** It should be above 20°C (68°F). Do not hesitate to cover your trays with a transparent lid to make the temperature and humidity higher.

MAKING COMPOST

Start composting now, and you'll have enough organic matter to mix into your soil in the spring – your vegetable garden will thank you for it!

BOKASHI COMPOSTING

With a bokashi bin, composting is simple and takes up very little space on your balcony. Instead of throwing away your organic waste, simply place it in your bokashi bin, and let it decompose. Always add a layer of bran on top of food scraps. Use the contents of the bin in two ways:

1. Every week, open the tap to release compost juice, a natural fertilizer for watering your plants. Mix 1 part of compost juice with 10 parts of water before using it.

2. When the bokashi bin is full, keep harvesting the compost juice every week for another two months. At the end of this period, bury the entire contents of the bokashi bin directly into the soil of your pots and planters. This will provide your soil with natural, nutrient-rich organic matter.

Saving time

You can easily buy a bokashi compost bin or worm bin online or at a garden centre.

WHAT CAN YOU PUT IN YOUR COMPOST?

You can put in all your fruit and vegetable waste, such as peelings, cores, and skins. You can also add crushed eggshells. However, do not put pasta, meat, or fish waste in your compost.

Saving money

You can make your own bokashi bin using a plastic bucket. All you need to do is poke a hole in the bucket to add a small tap for draining and collecting that valuable compost juice. You'll also need a lid to keep unpleasant smells in and promote decomposition.

COLLECTING RAINWATER

Even on a balcony, you can collect rainwater and reduce your water consumption. Every small sustainable gesture counts!

THE BENEFITS OF RAINWATER

The water you use to water your plants influences the soil's pH and affects plant growth. Most garden plants grow best with a neutral or slightly acidic pH, which is precisely the pH of rainwater!

WHAT ABOUT TAP WATER?

Sanitized drinking water generally has a pH of just over 7 (slightly alkaline), but in some cities, tap water can be very hard, with a pH over 8. Hard water is much less suitable for plants: if hard water is used in high doses, calcium deposits can form around roots, stopping them from absorbing the micronutrients from the soil that they need for growth.

Watering plants with the right water

Knowing the pH of your tap water is essential, so run a pH test. If your water is very hard, you'll need to alternately water your plants with rainwater and tap water. Otherwise, you can add half a squeezed lemon to a 10-litre (2.5 US gallon) watering can to water your plants. This will soften the water naturally.

Reusing water from your shower

When taking a shower, we often leave the water running until it gets hot enough. This is a huge waste, as the clean water will end up mixed with polluted water! To avoid this, put a bucket in your shower and let the water run into it until it reaches the desired temperature. You can then water your plants with it or store it in a barrel. This way, you can recover up to 80 litres (21 US gallons) of water per week for a two-person household.

RAINWATER HARVESTING IN A SMALL SPACE

If you have a rainwater gutter running over your balcony, you can tap into it and collect rainwater, provided you follow your apartment building's regulations. All you need to do is connect it to a small water barrel. Place the barrel near your plants if possible, to avoid having to go back and forth when watering.

Without access to a gutter, collecting rainwater will be more complicated. However, you can store reused water such as shower water (see above) on your balcony in a barrel with a lid.

SPRING

SPRING ON YOUR BALCONY

After the cold comes milder, warmer, and sunnier – though sometimes fickle – weather. In the vegetable garden, spring is a time of rebirth. The first rays of sunshine and longer days give the garden a new lease of life.

A LITTLE MORE PATIENCE

I love spring—in just a few weeks, nature will be back in full swing. From the first buds to full foliage, everything is suddenly alive again.

For a gardener, however, it's time to be patient. There's still a risk of frost, so don't rush! You'll have to bide your time: your seedlings still need time to grow, slowly but surely.

SPRING TO-DO LIST

- Choose pots, planters, and garden beds
- Select the right potting compost
- Care for seedlings and plant the first batch outdoors
- Keep filling your bokashi bin
- Prepare pots
- Plant seedlings into pots
- Spruce up your balcony
- Add mulch to pots
- Start fertilizing soil
- Watch out for pests
- Prune plants
- Plant flowers
- Harvest first crops
- Pollinate plants

SHOPPING AND REPURPOSING LIST

- New pots or reuse ones you have already
- Bags of clay pebbles or collect pebbles
- Potting compost
- Seed planting compost
- Collect straw, dry leaves, or grass clippings
- A 10-litre (2.5 US gallon) watering can
- Bamboo stakes
- Twine
- Small pair of pruning shears or kitchen shears
- Garden trowel or kitchen spoon
- Organic or homemade fertilizer

BE READY FOR HARVEST

Spring is also a key planning period to gather all the supplies you will need to create your own little urban jungle. This is the perfect time to prepare your pots, choose your potting compost, plant your first seedlings, and plant pretty flowers for our pollinating friends. And of course, you can look forward to your first harvests!

You will be able to enjoy the first crisp lettuces and crunchy radishes. Read the following pages carefully – joy is just around the corner!

CONSIDER YOUR CONTAINERS

To follow the Frenchie Gardener method, you'll need seven pots, five troughs or window boxes, and two raised planters. Here are some important factors to consider when choosing your containers!

MATERIALS AND COLOURS

You can choose between recycled plastic, terracotta, or geotextile. These materials all have their pros and cons, which I've listed in the table below to help you choose. It all depends on where you live (your climate and the type of balcony you have), your budget, and your preferences.

	Pros	Cons
Recycled plastic	– Very durable – Easy to store – Cheap – Light	– Not the most sustainable choice – Black plastic pots can store a lot of heat
Terracotta	– A natural, sustainable material – Useful for retaining water and coolness in the soil in the summer	– Can be pricey – Heavy – Not frost-resistant (might crack)
Geotextile	– Often made from recycled materials – Easy to store – Light to move	– Quite expensive

Saving Money

Remember there may be pots and planters at your local second-hand shop, where you can often find lots of gardening tools and supplies at very low prices too. You can also upcycle plastic containers to use as pots, as long as they are large enough for the volumes listed opposite. Large buckets, water bottles, large plastic tote bags, or crates will do the trick. Remember to poke holes in the bottom to allow excess water to drain away.

CAPACITY

This is the main factor to consider, as some cultivars need more soil to grow and flourish. You'll need to find a pot with a large enough capacity for the plant varieties listed here.

	Plant	Volume required
Pots	Basil	1 pot of 5 litres (1.5 US gallons)
	Cherry tomato	1 pot of 12 litres (3 US gallons)
	Aubergine, cucumber, pepper, potato, tomato	5 × 18-litre pots (5 US gallons)
Troughs	Nasturtium, strawberry, sorrel, phacelia, zinnia	5 × 15-litre troughs (4 US gallons)
Raised planters	Lettuce and radish	2 × 50- or 100-litre raised planters (13 or 26 US gallons), depending on available space

MATCHING POTS TO YOUR SPACE

When choosing, consider both your vegetable garden's surface area and fastenings on vertical surfaces for planters or hanging pots.

• **Pots** Allow floor space for these.

• **Troughs** These can be fastened to railings or walls, depending on your space. Make sure you use the right fastening systems for your railings. Inspect your walls to see whether you can drill holes in them or not, and whether they can bear the weight of pots filled with soil.

• **Raised planters** Measure the available length and width. I highly recommend raised garden beds to optimize your floor space.

COLOUR

There is a huge choice of colours for containers. You can select them to match your surroundings, to create a soothing space, or to stand out and make a colourful statement.

Saving time

If you're looking to buy pots, troughs, or raised garden beds, search online or visit a garden centre.

MAKING UPCYCLED WALL POTS

Upcycle 1-litre (about 0.25 US gallon) HDPE plastic bottles and repurpose them as herb pots – for mint, oregano, basil, and many others:

1. Cut out the bottom.

2. Poke holes into the cap to allow the water to drain; to make it easier, you can heat up a needle or a small screwdriver over a candle to punch the holes into a plastic cap.

3. Hang the bottle with the cap down, either by drilling to screw in the bottle or using twine, depending on your wall.

4. Pour in a thin layer of clay pebbles, add potting soil, and plant your seedlings (only one per bottle!).

QUALITY POTTING COMPOST

The key to gardening success is quality potting compost — and it will help to heal your basil trauma, too (see page 6). You do have green fingers after all...

IT'S ALL ABOUT THE SOIL!

Don't give in to the temptation of cheap multi-purpose potting compost! While you can afford to save money on pots and fertilizers, quality soil is a necessity for a healthy, fertile vegetable garden.

To grow, a plant needs light for photosynthesis, water, and nutrients from the soil. Begin with a quality potting compost to offer it the best possible start.

SO WHICH ONE DO I NEED?

• Avoid supermarkets and multi-purpose potting composts; buy your soil at a garden centre.
• Whatever soil you buy, make sure it is organic, pesticide-free potting compost.
• Choose potting compost specifically for growing fruit and vegetables: it will provide the required nutrients to make sure your plants get off to a healthy start.
• Check whether it can be used in organic growing by looking closely at the packaging.
• If it's already enriched with organic compost, all the better!
• Potting compost with perlite is best. Perlite is a natural volcanic rock that improves soil drainage and fluffiness. It is also sold separately and can be added to your soil.

About peat

Peat is partly decomposed plant matter used to enrich the soil. Natural peat bogs significantly reduce global warming by storing carbon. So avoid using peat and replace it with quality compost. If you buy organic soil, make sure it's peat-free.

Potting compost is not enough on its own!

• **Potting compost** This is the foundation of your vegetable garden and what nourishes your plants.

• **An additive** This is an optional component that enriches potting compost to make it even more nutritious. For instance, worm castings.

• **Put them together** You'll need a mix of potting compost and additive to suit your budget and what you plan to plant.

GARDEN MATHEMATICS!

Once you have selected your potting compost, you need to work out the total amount for all your pots. To do this, find each pot's capacity and add them all together to get the total soil volume. This will tell you how much potting compost you need. Potting compost and additives are usually sold in 25- or 50-litre bags (20 or 40 quart bags in the US).

To follow the Frenchie Gardener method, you'll need a total of 282 litres (297 quarts) (see table). I recommend the following breakdown: two parts potting compost and one part additive. For 50-litre bags, that means four bags of potting soil and two bags of additive, such as worm castings.

	Volume	Total
Pots	1 pot of 5 litres (5 quarts)	5 litres (5 quarts)
	1 pot of 12 litres (12.5 quarts)	12 litres (12 quarts)
	5 × 18-litre pots (19 US quarts)	90 litres (95 quarts)
Troughs	5 × 15-litre troughs (16 quarts)	75 litres (80 quarts)
Raised planters	2 × 50 or 1 × 100-litre raised planters (105 quarts)	100 litres (105 quarts)

Saving money

You can reuse the potting soil from the previous year to save on a few bags of potting soil, provided it is neither mouldy nor too pebbly. To start a new pot, I recommend a mix of one-third old potting soil, one-third new potting compost, and one-third compost. The rest is up to you!

Although clay pebbles are best, you can use tiny stones to put in the bottom of your pots instead.

STRAW AND CLAY PEBBLES

In addition to potting soil and substrate, you'll need a 50-litre (50 quart) bag of clay pebbles to line the bottom of your pots and create a thin layer between your soil and the drainage holes. The pebbles will help keep your soil fluffy and drain water.

You'll also need straw (see page 66), which you can find at garden centres or pet shops. A 25-litre (26 quart) bag will do.

PLANTING SEEDLINGS INDOORS

Seedlings planted over four weeks ago need to be planted into slightly larger pots in which they can continue to grow indoors for a while.

WHICH SEEDLINGS?

You'll need to plant tomatoes, aubergines, peppers, and basil into a slightly larger pot, as they've been drawing nutrients from the seed compost for over four weeks. It's time to give them new, nutrient-rich potting compost and more room to root. To do this, plant them into the potting compost you'll be using for your outdoor pots.

PLANTING THE SEEDLINGS

This is an important but delicate stage, as the seedlings are still fragile. They must be handled with care.

1. Take a half-litre (0.1 US gallon) or upcycled pot and fill it with potting compost.

2. Water and let the water drain.

3. Press down the soil to settle it.

4. Make a hole the size and depth of one section of the seed tray (or other pot) in which you sprouted your seedling.

5. Press lightly on the sides and under the pot containing the seedling to loosen the compost.

6. Use the fork technique: slide a fork between the soil and one side of the pot. Delicately lever out the seedling and its soil.

7. You should have a seedling with undamaged roots. Hold it by the rootball.

8. Place the seedling and its rootball in its new pot.

9. Pack the soil all around and press down lightly. Add potting compost if necessary.

10. Water carefully.

11. Label your seedlings.

12. Place the seedlings about 15cm (6in) from your grow lamp or set them up on a well-lit windowsill. Continue watering daily as soon as the original soil starts to dry up.

If you are 4 weeks away from your last frost date...

This is a good time to start your cucumber, sorrel, and lettuce seedlings, which will then have time to sprout properly before being planted into the ground after your last frost date. No additional potting required for these!

PREPPING POTS AND PLANTERS

Before planting seedlings outdoors, I recommend that you prepare your soil around two weeks in advance, so that it begins to work to welcome your seedlings into ideal conditions.

FILLING POTS AND WINDOW BOXES

1. If your pots are new, proceed to step 2. Otherwise, start by washing the pots with organic soap and water to remove any bacteria or pest eggs.

2. Add a layer of clay or stone pebbles to the bottom. A thin layer is sufficient: you only need enough to line the bottom of the pot.

3. Add 3–4cm (1¼–1½in) of straw, then cover with a thin layer of potting compost (you can reuse old potting soil as mentioned previously).

4. Add organic matter, a key component in your pot's ecosystem. Pour in some coffee grounds at a ratio of one part coffee grounds to 20 parts soil.

5. You can also add ground (ideally powdered) eggshells to speed up decomposition.

6. Using a garden trowel, scoop the most decomposed material from the bottom of your bokashi composter. Pour two heaped scoops into your pot.

7. Now add about two parts potting compost and one part additive (such as worm castings), mixing as you go. Stop 5cm (2in) from the rim of the pot.

8. Water well and let the water drain.

The soil level will probably drop once the water has drained: add potting soil and compost to raise the soil level to within 5cm (2in) of the pot rim.

Let the soil settle: it will start to assimilate the organic matter. In a week or two, it will be ready for your plants!

FILLING RAISED PLANTERS

Follow the same steps as for pots and window boxes, but using larger volumes. For a raised planter that is about 50 litres (50 quarts) in volume, adjust quantities accordingly. I find that about six heaped scoops of bokashi compost works well.

In the second step, if your planter has hollow legs with drainage holes, pour the clay pebbles or stones into the legs, filling them up to the top.

PLANTING SEEDLINGS OUTDOORS

Is your last frost date just around the corner? I bet you're excited about setting your plants outside on your balcony. However, the weather can be unpredictable, and it's best to wait a few more days to be sure it's safe to plant outdoors.

BEFORE TAKING THE PLUNGE

Take a look at the weather forecasts for the next two weeks. If there's no risk of frost, you can start planting certain seedlings outdoors, such as lettuces and sorrel. You can also plant radishes and potatoes outdoors (see page 63). Let's get to work!

Ice Saints

In France and Eastern Europe, there is believed to be a period of cold weather around the feast days of St Mamertus, St Pancras, and St Servatius, celebrated on 11, 12, and 13 May respectively. They are known as the Ice Saints. Since the Middle Ages, almanacs have warned farmers and gardeners to look out for a frost that could negatively impact crops around this time of year. In North America, indigenous peoples have referred to this time of year as the full flower moon, milk moon, or corn planting moon. It's pretty scary if you look at your last frost date, as there may still be a late frost that could put your plants at risk... However, statistically it's still very unlikely to have frost at this time of year: so you can rest easy!

PLANT LETTUCES AND SORREL

Easy peasy – all you have to do is gently remove the seedlings from their pots (see pages 54–55). To do this, it's best if the soil is not wet, so the root ball will be easier to remove from its container.

Once you have removed your seedlings, plant eight lettuce seedlings in a raised planter, spacing them evenly. For sorrel, plant three plants in one of the raised garden beds, again giving each seedling sufficient space (check the packet for the ideal spacing).

After planting, pack the soil around your seedlings and water the soil around them, so that the original soil in your planter blends with the soil containing the rootball of each seedling.

PLANT TOMATOES, AUBERGINES, CUCUMBERS, AND PEPPERS

Even if the last frost date has passed, wait for the temperature to consistently be above 10°C (50°F) at night before planting these species outside. Don't forget that these are summer fruits and vegetables, which need sunshine to thrive. So there's no need to rush!

• For cherry tomatoes, use a 12-litre pot (3 US gallons). For aubergines, cucumbers, peppers, and larger tomatoes, you'll need an 18-litre pot (5 US gallons).
• Before you start planting, make sure you have a bamboo stake and some string to hand (see pages 60–61).
• For easy planting, water your soil and use the pot moulding technique (see box below).

Easy planting: the pot moulding technique

To be sure you are making the right size hole in the planting pot, water the potting soil generously, then press down using the container in which your seedling is growing. This will give you the perfect size (depth, width, and height) to accommodate your seedling and its root ball (see page 60). Simple, right?

5. Carefully turn your seedling right side up, holding the stem upright.

6. Place your seedling in the hole.

7. Pack well around it, then water at the base.

8. Now insert a bamboo stake at an angle in your pot, so as not to damage your plant's roots.

9. Tie the stem to the stake with string to support it. Don't tie it too tightly, as the twine could soon damage it when your seedling grows.

Bury tomatoes deeply

For tomatoes, you can use the pot moulding technique, but you'll need to dig a little deeper so the seedlings are buried up to the first two small leaves on the stem. This will enable your tomato plant to grow more roots to anchor it firmly in the soil.

PLANTING SEEDLINGS INTO THEIR FINAL POTS

This is a delicate process, as their stems are still fragile and need to be handled with great care.

1. Mould the soil around the seedling pot by pressing it firmly into the larger container.

2. You should obtain a perfectly shaped hole.

3. Slide your fingers around the stem of your seedling and turn it upside down.

4. Squeeze the pot so the soil and roots come out.

PLANTING STRAWBERRY PLANTS

Plant two strawberry plants into a planter using the pot-moulding technique (see page 59). If you've never sown strawberry plants before, go to a garden centre to buy seedlings. Planting is very simple, as long as you handle the plant with care.

PLANTING BASIL

Plant in the same way as lettuce (see page 58), in a 5-litre pot (1.5 US gallons). Firm the soil well around the plant and water it. You many want to add a straw mulch (see page 66). Keep an eye on nighttime temperatures, however, as basil needs warmth and sunlight to thrive. If temperatures fall below 10°C (50°F), keep your basil indoors, near a sunny window, to give it time to grow.

PLANTING DIRECTLY IN THE GROUND

There's no need to grow seedlings for radishes and potatoes: you can plant them directly in the ground after the last frost date.

PLANTING RADISHES

1. Using a pencil, make holes 1cm (½in) deep and 5cm (2in) apart all over the surface of one of your raised garden planters. You should be able to fit in around 40 holes.

2. Place one seed in each hole, then carefully cover them. Firm the potting soil with your hands, then lightly water your raised garden planter table by misting with a spray bottle, so the seeds don't rise to the surface and come out of their holes.

3. The first shoots will appear within a few days. Keep the soil moist to ensure germination.

PLANTING POTATOES

1. In an 18-litre pot (5 US gallons), make 2 holes with a broom handle or neck of a glass bottle. The hole should reach half the depth of the pot (or you could fill only half the pot with potting soil). Place one sprouting potato per hole, then cover. Firm the soil and water generously.

2. Green stems will soon sprout from the soil and produce beautiful leaves. In a few months, when the leaves dry and wilt, you will be ready to harvest your potatoes.

ORGANIZING YOUR BALCONY FOR SPRING

Now that everything's planted, you need to position your pots, troughs, and raised garden planters on your balcony.

AN IMPORTANT STEP

Organizing a balcony is important for two reasons:
• Optimizing space and making sure it's easy to move around your balcony.
• Providing the different plants you'll be growing with the right amount of sunlight.

The perk of growing in pots is that you can easily move them around and adapt them to your needs, and those of the plants! This means you can readjust or swap your pots' locations as the seasons change. Your balcony is a flexible space, so make the most of it.

PAY ATTENTION TO THE WIND

What direction does it usually blow from? Is your balcony sheltered from the wind or is it exposed? Strong gusts could damage your plants, or even break the main stem. If there is a lot of wind around your home, place your pots near the balcony railings and securely fasten the stakes to the railings with a cable tie or twine.

Ideal exposure

In early spring, temperatures are still cool, especially at night – so it's important to expose cultivars that need sunshine to as much of it as possible in the day.
Place in direct sunlight: radishes, tomatoes, aubergines, cucumbers, peppers, strawberries, basil, zinnias.
Place in semi-shade: lettuces, sorrel, nasturtiums, potatoes, phacelia.

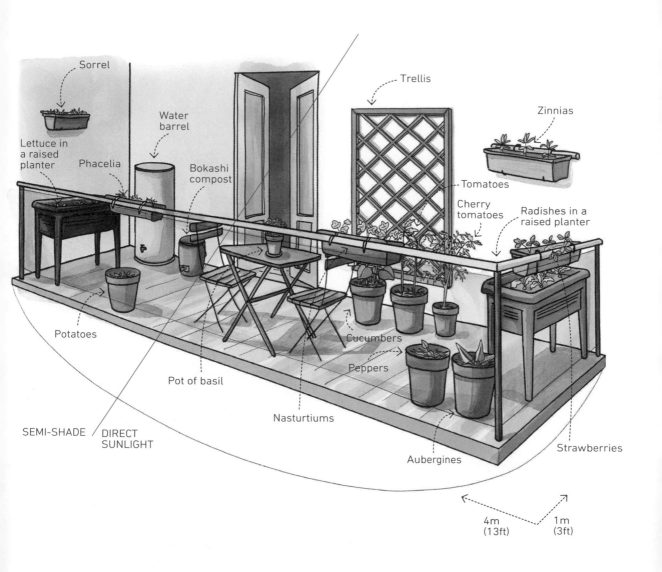

Sorrel

Lettuce in
a raised
planter

Phacelia

Water
barrel

Bokashi
compost

Trellis

Zinnias

Tomatoes

Cherry
tomatoes

Radishes in a
raised planter

Potatoes

Pot of basil

Nasturtiums

Cucumbers

Peppers

Aubergines

Strawberries

SEMI-SHADE / DIRECT
SUNLIGHT

4m
(13ft)

1m
(3ft)

MULCHING

"Why do you put straw in your pots?" The answer is simple: it's not just for looks, it's ultra-beneficial for the soil and the plants!

A GARDENER'S SECRET

Mulching involves adding organic matter to the soil in your pots. Mulching has many benefits:
• **Improved water retention** Mulching promotes cooler, wetter soil conditions in summer. Water will evaporate less, so the plant will have more water to absorb.
• **Insulation** In autumn or winter, mulch will protect your plants from the cold by acting as an insulating layer.
• **Enriched soil** As it gradually decomposes, a straw mulch nourishes your soil with nutrients. Simple and effective.

Saving money

Do you live near a farm that has extra straw? Go ahead and ask – you just might have found an affordable supplier. If you live in a big city, you can find alternative mulches for free:
Dry leaves If you see dry leaves on the pavement, grab a bag to collect them. They'll make terrific mulch. If the leaves are on grass or in the forest, don't pick them up, as they'll feed the soil where they have landed.
Freshly cut grass If your neighbour or your council mows the lawn in a small garden or a large park, and leaves the cuttings on the surface, ask if you can collect the freshly cut grass.

Saving time

Buy straw from a garden centre or pet shop – it's perfect for your plants!

ON YOUR MARKS, GET SET, MULCH!

You can mulch your pots right away, adding a layer about 3cm (1in) thick on top of the soil in your pots. Your plants will thank you, and so will the planet as you are saving water!

WATERING YOUR VEGETABLE GARDEN

For some, watering is a chore. For me, it's a moment to relax and even meditate. It allows me to take a break and connect with nature, smell the soothing scent of wet soil, and check in with my plants.

DIFFERENT NEEDS

Watering varies depending on the plant type and the weather. The table below will help you to water each plant accordingly:

Low water requirements	Moderate water requirements	High water requirements
Basil	Pepper, lettuce, radish, sorrel, aubergine	Tomato, potato, cucumber

TWO GOLDEN RULES

• **Water early in the morning or at sunset** This timing will allow the plants to absorb water better and prevent it from evaporating quickly. It is not recommended to water in the middle of the day, and even less so in direct sunlight. A single drop of water on a leaf can magnify sunlight and burn the plant.

• **Water at the base of the stem and avoid splashing the leaves** Drops of water on the leaves will probably fall outside the pot, which is a waste of water. What's more, some plants, such as tomatoes, can be more likely to develop diseases if their leaves are wet.

UPCYCLING TIPS

• **Place saucers under your pots to collect excess water** Use it to water another plant, or store it in your watering can or water barrel.

• **Keep your cooking water** If you boil vegetables without adding salt, this water is perfectly reusable for your plants, provided you let it cool first. What's more, it contains some nutrients from the vegetables you've cooked!

• **Place a bucket in your shower** Collect the clean water you let run at the start of the shower until it reaches the desired temperature (see page 43).

FEEDING YOUR PLANTS

Most vegetable garden failures, especially in containers, are due to insufficient feeding. To ensure a bountiful harvest, remember that a plant needs the right nutrients at the right time to grow well. Since your soil's nutrient content will deplete over time, feeding your vegetable garden is essential!

THE HOLY TRINITY OF NUTRIENTS

Vegetable plants mainly need three macronutrients: nitrogen, phosphorus, and potassium.
- **Nitrogen (N)** This is an essential nutrient for plant growth: it boosts stem, root, and leaf growth.
- **Phosphorus (P)** This is important for plants with a long growth cycle, which need to flower before producing fruit. Phosphorus ensures better flowering and fruiting.
- **Potassium (K)** This is needed to generally keep a plant healthy. It supports the plant's immune system, ensuring healthy growth and resistance to disease.

NUTRITION Á LA CARTE

Depending on the species and the growth stage of the plant, you will need to provide a certain amount of macronutrients, especially phosphorus. You will thus need to adjust your feeding according to the species you are growing.
- **Short growth cycle** Plants such as lettuce, radish, basil, sorrel, and potatoes have a cycle of sowing, growth, harvest, flowering. They don't need to produce flowers to crop; you will in fact harvest them before they flower. Since you will be eating the leaves (or roots), you'll need to provide them with a regular supply of nitrogen and potassium. Phosphorus will not be vital.

SHORT GROWTH CYCLE
e.g. lettuce

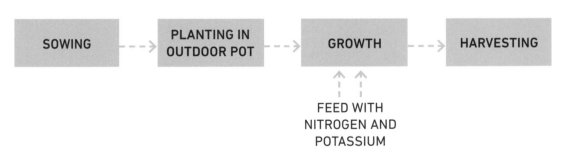

• **Long growth cycle** Tomato, aubergine, cucumber, and pepper have a cycle consisting of sowing, development, flowering, harvesting. These plants will need nitrogen and potassium for their growth phase, to stimulate root, leaf, and stem development. When the first flowers appear, you'll need to cut back on nitrogen and potassium and give them more phosphorus to encourage beautiful flowering and healthy fruiting.

Finding the right fertilizer

Buy pesticide-free organic fertilizers. I strongly recommend seaweed-based fertilizers, which provide nitrogen and potassium. As for phosphorus, guano works wonders! Also consider worm castings, which is a perfect balanced organic fertilizer. Each fertilizer has a measurement number on the bottle, with the abbreviation NPK: N for Nitrogen, P for Phosphorus, and K for Kalium (potassium). For example, 4-2-4 (N = 4 parts, P = 2 parts, K = 4 parts) means the fertilizer contains more nitrogen and potassium than phosphorus. Conversely, you may also find fertilizers with 2-8-2 (N = 2 parts, P = 8 parts, K = 2 parts), where the phosphorus concentration is higher.

LONG GROWTH CYCLE e.g. tomato

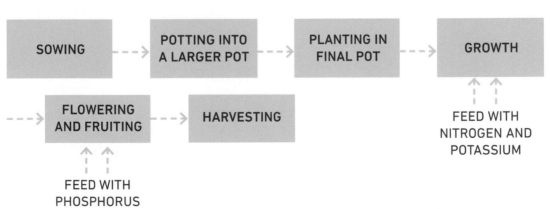

SOWING → POTTING INTO A LARGER POT → PLANTING IN FINAL POT → GROWTH

FLOWERING AND FRUITING → HARVESTING

FEED WITH NITROGEN AND POTASSIUM

FEED WITH PHOSPHORUS

MAKING YOUR OWN FERTILIZER

Instead of buying fertilizers, you can make your own by reusing plant-based waste. They're really effective, and complement the compost juice in your bokashi. Here are the easiest to make at home.

COFFEE GROUNDS FOR NITROGEN

Don't throw away your coffee grounds – they're very valuable for your soil and plants. You can add them to your compost or mix them directly with potting soil. Earthworms love them.

BANANA WATER FOR PLENTY OF POTASSIUM AND PHOSPHORUS

Treasure your banana peelings! Simply soak a banana skin in a 2-litre container (0.5 US gallon) for two days, then water your plants with it.

EGGSHELLS FOR MINERALS AND CALCIUM

If you love making omelettes, keep your eggshells. Grind them into small pieces with a pestle and mortar or put them through a blender to powder them. All you have to do next is add them to your potting soil.

YOUR URINE FOR NITROGEN

Yes, I am serious! Human urine is a free and effective nitrogen fertilizer. Mix 1 part urine with 10 parts water, then water directly into your pots. Keep the urine for 24 hours at the very most and stop fertilizing with urine about one month before harvesting. It's also not recommended to use urine as fertilizer during menstruation or if you're taking antibiotics. Otherwise, it works like a charm!

FEEDING IN SPRING

In the spring, you can start fertilizing four weeks after planting in your outdoor pots. Feel free to fertilize using your compost juice once a week.

A DILIGENT GARDENER'S PLAN

• **Four weeks after planting out** All your plants need is a little nitrogen and potassium to kick-start their growth phase, helping them to sprout attractive roots and foliage. For this, I highly recommend using nettle fertilizer in early spring (see below).

• **After four weeks** Fertilize your plants every two to three weeks at most, depending on their needs. Bear in mind that fertilizing too regularly can damage your soil and crops. It's all a question of balance and patience!

• **First flowers** When flowers first appear on your tomato, aubergine, pepper, and cucumber plants, don't forget to use high-phosphorus fertilizer to boost plant growth. It works wonders!

HOMEMADE MIRACLE NETTLE MANURE

Nettle is a fantastic plant ... when it doesn't sting. By fermenting it, you can make a powerful and very useful fertilizer to boost the immune system of young plants in early spring. Be careful, however, nettle manure doesn't smell very nice – sensitive noses be warned!

1. Harvest some nettles in a forest or park. Don't harvest near a road or in town: they will have absorbed pollution from exhaust fumes. Cut nettles off at the base of the stem to encourage regrowth. And wear gloves to avoid stings!

2. When you get home, chop the leaves into small pieces and place in a 1-litre (about 34 fl oz) jar or bottle.

3. Pour 1 litre (about 34 fl oz) of water into the container (preferably rainwater, but if you use tap water, leave it to stand for 24 hours beforehand).

4. Close the container and shake it well.

5. Open the cap to allow a little air in.

6. Repeat steps 4 and 5 every day. Fermentation bubbles will start to appear – a good sign! Do this for about 10 to 15 days. When you see no more bubbles, your fertilizer is ready!

7. Collect the liquid using a sieve over a large jug or bowl. Put the leaves in your bokashi bin.

8. In a 10-litre (2.5 US gallon) watering can, mix 1 litre (0.25 US gallon) of nettle fertilizer with 9 litres (2.25 US gallons) of water, and water your plants with it.

SPRING PESTS

When your plants start to grow, certain insects and other creatures will also want a piece of the action...

APHIDS

At the beginning of spring, the aphid is the most common pest. Green, black, brown, grey, even red–whatever its colour, it doesn't make it any cuter! This little insect generally settles under plant leaves and feeds on their sap. Nothing serious if you detect them in time – inspect your plants regularly, and don't forget your seedlings. Another indicator can be the presence of ants on your plants. Ants domesticate aphids in order to collect the honeydew they produce, a sweet liquid that the ants can't get enough of!

Getting rid of aphids

Simply crush them with your fingers or use an old soft-bristled toothbrush to brush the leaves to remove aphids and their eggs. If the colony is already fairly large, mix 1 teaspoon of organic soap in 1-litre (about 34fl oz) of water. Spray the mix onto the plant twice a week. This is a natural insecticide that won't harm plants.

SLUGS AND SNAILS

Slugs and snails can wreak havoc on your plants, especially when they've just been planted. You'll only see these plant predators at night, when it's damp. So grab a torch and look around. If you see holes in some of the leaves, slugs and snails can't be too far away. Fortunately, there are simple and effective solutions to counter the invasion (see box below).

Slug and snail control

• **Cover seedlings** Repurpose a transparent plastic container by cutting small holes for ventilation, and placing it over your seedlings.
• **Protect the stem** Build a natural barrier of broken eggshells on the soil around each stem. Slugs and snails don't like moving over dry, rough surfaces.
• **In case of an invasion** Pour beer into a bowl. Slugs and snails love the smell and will flock to it. All you have to do now is capture them (don't hesitate to hold them to ransom)!

PRUNING IN SPRING

To ensure they start to grow well, some types of vegetable plants will need pruning to help them develop a strong, productive shape. For pruning later in the summer, see page 92.

TOMATOES

• **Prune the leaves at the base** Once your tomato plant has developed dense foliage of around seven or eight branches and has grown to 60–70cm tall (24–28in), cut the lower branches back to the main stem, especially if the leaves are touching the ground.

Tomatoes hate humidity and having wet leaves. This moisture can lead them to develop mildew, a fungus that infects the green parts of a plant, often causing the plant's death. By removing the first two or three branches at the bottom, you reduce the risk of getting water on the leaves while watering. Pruning these leaves also allows the plant to keep growing taller. So go ahead, your tomato plant will thank you!

• **What to do with side shoots** These are smaller shoots that grow between the main stem and a branch, at a 45-degree angle. They absorb a lot of nutrients and take energy away from the main stem, slowing its upward growth. Whether or not to remove them is a huge debate in the gardening community. First of all, you need to check whether the tomatoes you're growing are determinate or indeterminate in the way they grow and produce their shoots (see page 74).

What to do with pruned side shoots?

- Put them in your pot as mulch.
- Add them to your bokashi compost.
- Place your shoot in a glass of water. After a few days, new roots will appear: you now have a free, young tomato plant!
- You can make natural, homemade fertilizer (see below, right).

• **Determinate tomatoes** These can grow to around 1.5–2m (5–6½ft) tall, and will produce tomatoes in a single batch. In this case, removing the side shoots means cutting off a stem that could have produced more tomatoes, so this is not advised.

• **Indeterminate tomatoes** These can reach heights of over 2m (6½ft) and produce continuously throughout their life cycle. Removing the side shoots encourages the plant to grow even taller. But is this really what you want on a small balcony? In my garden, I cut them off on the first five or six branches, then stop. I find it is best to cut these out only at the beginning of the plant's growth phase, before it starts producing tomatoes. It's up to you to experiment!

Simply cut side shoots off with clean secateurs to avoid spreading any disease or bacteria from one plant to the next.

Make homemade fertilizer using side shoots

Here's a simple and effective recipe for fertilizing your tomatoes. Before using it, make sure the plants are healthy and disease-free (see pages 94–97).
1. Chop 100–150g (3½–5oz) of tomato side shoots into small pieces and place in an empty 1-litre (34fl oz) bottle.
2. Fill the bottle with water, preferably rainwater.
3. Close the cap, but not completely, so that air can flow slightly.
4. Let it sit and ferment for 4 days, stirring daily. To do this, close the cap completely and shake the bottle gently up and down. Open the cap again to let some air through.
5. After 4 days, strain the mixture through a colander and collect the liquid. The fermented leaves can be composted.
6. Dilute your tomato fertilizer in 3 litres (100fl oz) of water, and you've got a homemade fertilizer for your tomatoes!

AUBERGINES AND PEPPERS

To prune aubergines, use the same technique as for tomatoes: cut off the lower leaves and the first side shoots. In spring, your aubergines and peppers may be at a much less advanced stage of growth than tomatoes, which is perfectly normal. As soon as temperatures start to rise, you'll see them grow drastically in just a few weeks.

BASIL

Basil plants need to be pruned when they reach 20–30cm (8–12in) tall. Off with their heads! Cut off the top of each plant to stimulate new shoots at the joint of each leaf below. So, instead of one main stem, you'll get two or even four new stems, which will each sprout new leaves. This technique also works for mint. See page 90 for how to make new basil plants with the growth you have cut off.

STRAWBERRY PLANTS

• **Plants grown from seed** You'll probably notice that the plants are still compact, with just a few small leaves. Don't prune them at all in their first year, but let them continue to grow so that their roots and foliage can develop. They won't produce fruits in the first year, but will grow into strong plants.

• **Plants grown on from seedlings** Their foliage will already be more developed, and you may even have one or two flowers, which you'll need to cut off to tell the plant to keep growing. This is the only pruning you'll need to do, but it stimulates the plant to keep developing its roots and foliage. The more patient you are, the healthier your plants will grow, and the more strawberries they'll produce.

So cut off those first flowers – this will help you to have a beautiful harvest later in the year! For making new plants from strawberry plants in summer, see page 91.

DON'T PRUNE POTATOES AND CUCUMBERS!

• **Potatoes** Simply let the foliage grow as much as you like. Enjoy the greenery.

• **Support vertical growth of cucumbers** You'll probably find little green stalks growing out of the side of the plant. These tendrils are looking for support to make sure the plant can keep growing upwards. If you point these stems towards the top of your bamboo stake, they will slowly curl around it for support.

POLLINATING VEGETABLE PLANTS

To produce fruit and vegetables, the flowers on your plants need to be pollinated. Bees, bumblebees, and other pollinators should be doing their job, but if you don't see enough of them, it's up to you.

THREE SIMPLE METHODS

Tomatoes, peppers, and aubergines contain both male and female parts on the same flower. Simply stimulate the flower to mix pollen from the male part, the stamen, with the female part, the stigma. You can choose between three pollinating methods:
• Gently tap the back of the flower with your finger to shake it.
• Gently tickle the inside of the flower with a paint brush.
• Gently shake the main stem of your plant to recreate the effect of wind.

Help, my plant's flowers are falling off!

If it's just a single flower, it's no big deal. On the other hand, if a lot of flowers are falling off, you need to act fast. This may be a sign of a lack of water, so water your plants regularly. You can also give them a boost of phosphorus using guano (see page 69).

What about the rest?

For your lettuces, radishes, sorrel, basil, and potatoes, there's no need to pollinate as you will harvest them before they flower. In fact, I bet you'll soon be enjoying your first harvests!

POLLINATING CUCUMBERS

Unlike tomatoes, cucumbers have a male flower containing the pollen, and a female flower containing the pistil (including the stigma). For successful pollination, a bee visits the male flower first, collecting pollen, then visits the female flower, depositing pollen on the pistil.

The cucumber grows out of the female flower. The female cucumber flower has a thicker stem, which looks like a mini gherkin.

To make sure you get cucumbers, take a small paint brush, and gently transfer pollen from the male flower into the female flower.

FLOWER POWER

Generally speaking, the more flowers you have, the healthier and more productive your garden. Given our limited space, the challenge is to select the flowers that will create the best possible ecosystem.

ATTRACTING BEES

Bees play an important part in the vegetable garden. They feed on the nectar from the flowers of your vegetable plants and pollinate them, raising your chances of obtaining fruit and vegetables. To attract them, you can plant lavender, blueberries, or plants with daisy-type flowers, all of which are a big hit with bees. My favourite, however, is phacelia, which bees love. What's more, it will self seed and grow again by itself the following year, so what more could you ask for?

TRAPPING PESTS

To attract aphids to a single spot in your vegetable garden, plant nasturtiums, which they love! They'll leave your other plants alone. It's a simple and effective way of controlling aphids without the need for chemical pesticides.

Ladybirds can also do the job, devouring up to 100 aphids a day. By planting nasturtiums, you're offering the best possible staycation: they'll be sure to give you a five-star review!

ATTRACTING FRIENDLY INSECTS

Even in the city, flowers contribute to biodiversity. In addition to ladybirds, flowers also attract bumblebees, butterflies, grasshoppers, and more. Zinnias will be a splendid addition to your balcony and will encourage these beneficial insects to visit. I bet you will love this beautiful, versatile flower – it will spruce up your exterior, protect biodiversity, and you will even be able to make beautiful bouquets!

PLANTING FLOWERS

Flowers are your best allies in the vegetable garden, attracting pollinators and effectively repelling certain pests. So on your marks, get set, plant!

PLANTING ZINNIAS

Zinnias (below) are sown directly into the ground. Take a windowbox or trough, water the soil, then make a first hole 2cm (a little less than 1in) deep, and a second 10cm (4in) away from the first. Place one seed in each hole, fill in, then water lightly. Keep the soil moist, and in a few days, the first shoots will appear above the soil.

SOW PHACELIA

Broadcast around 30 seeds in your window box or trough, cover with a thin layer – about 1cm (½in) – of potting soil, and water lightly. They'll soon be sprouting!

PLANTING NASTURTIUMS

Planting nasturtiums is very simple: follow the same method as for strawberries or lettuces (see pages 58 and 62). Plant four nasturtium seedlings (above) in a trough or windowbox.

As well as being beautiful, these flowers will be one of your best allies in attracting aphids. An ideal and pretty trap!

YOUR FIRST VEGETABLE CROPS

The short-cycle vegetables have surely grown well by now, and you're about to enjoy the first harvests. How and when to harvest? That depends on the species!

LETTUCE

For leafy vegetables like lettuce or sorrel, start harvesting leaf by leaf. By picking only the outer leaves, you'll get just the right amount for your needs, and the core will continue to grow and produce more leaves. Be careful, though: lettuces should be picked before they start to bloom. If a lettuce starts to grow upwards and you see a stalk forming, it's time to harvest it completely, as it will lose its flavour.

You can also start new lettuce seedlings, this time outdoors, to ensure you have a good supply of leaves through the summer. The lettuce seedlings will be ready for planting within three weeks.

HERE'S TO CRUNCHY RADISHES!

It's important to harvest radishes at the right time, especially if temperatures are rising: the longer you wait, the more bitter they'll be. On the other hand, radishes can also bloom. So take note of the maturity stage of the radish cultivar you're growing to know if it's ready to eat. To do this you can check the packet, or pick a radish to test its size and taste.

Once the radishes have been harvested, scrape the top of the potting soil off the raised garden bed to remove the roots. Bury some bokashi compost, and all you have to do is plant new seeds for the next harvest, which will be in one month tops!

Harvesting basil

Continue to chop off their heads as described on page 75. New leaves will keep growing!

SUMMER

SUMMER ON YOUR BALCONY

It's finally summer! The days are getting longer and brighter. The sun rises even higher in the sky, sending down warm rays for long hours. It's time to enjoy a picnic at the park, an aperitif al fresco, or even on your vegetable garden balcony.

MAKE THE MOST OF SUMMER IN YOUR VEGETABLE GARDEN

It's the perfect time to enjoy aperitifs and dinners on your balcony, garnished with freshly harvested fruit and vegetables. Radishes, tomatoes, lettuces, strawberries, and peppers will all be on the menu. Use fragrant basil to enhance your smoothies and other fresh juices.

You can also garden much later into the evening, and admire how this urban green jungle thrives and grows day by day.

SHOPPING AND REPURPOSING LIST

• Fertilizer with a high phosphorus content (such as guano)
• Sturdy stakes to support your plants
• Organic soap for pest attacks
• Seed compost if you run out
• Seeds for autumn

SUMMER TO-DO LIST

• Organize your balcony
• Fertilize the soil
• Harvest crops at the right time
• Put up supports for plants that need them
• Prune vegetable plants
• Pinch out flowers
• Inspect plants for pests and diseases
• Propagate basil and strawberry plants
• Water regularly
• Start autumn seedlings
• Rotate crops
• If you're going on holiday, stock up on empty plastic bottles, or buy irrigation pots or an automatic watering system

DON'T REST ON YOUR LAURELS

Summer is the season when you need to water regularly and watch out for heatwaves and thunderstorms. It would be a pity to ruin all your efforts by a moment of carelessness.

Maybe you're on your summer break too... But what about your vegetable garden if you're going away? Don't worry, I've got you covered.

PREPARING FOR AUTUMN

During the summer, you'll start preparing for autumn and organizing crop rotation. In gardening, planning is just as important as the sheer joy and wonder of growing. Have a great summer in the garden!

ORGANIZING YOUR BALCONY FOR SUMMER

Your balcony's summer layout will not be drastically different from its spring layout. Some of your plants have grown taller, so it's best not to move them around unnecessarily to avoid damaging them.

OUT OF THE SHADOWS...

With the much warmer summer temperatures, feel free to move your radish raised planter to a shadier spot by your lettuces. With the heat and direct sunlight, these cultivars will go to seed even faster. The same applies to sorrel.

Place your water barrel in the shade to reduce evaporation. Finally, your bokashi compost will also be more at home in a shady area.

... AND INTO THE LIGHT

Your potatoes prefer direct sunlight. Their foliage is now well formed, so the plant will continue to grow.

What to do in the event of a heatwave?

A heatwave can cause serious damage to plants, or even kill them. The flowers might wilt in the sun, affecting production. The leaves of some plants may curl up to protect themselves.

Feel free to move pots that are easy to move around into the shade during the hottest hours of the day to protect them. For plants that can't be moved, such as cucumbers or tomatoes, put up a canopy to shield them from the sun and provide some shade.

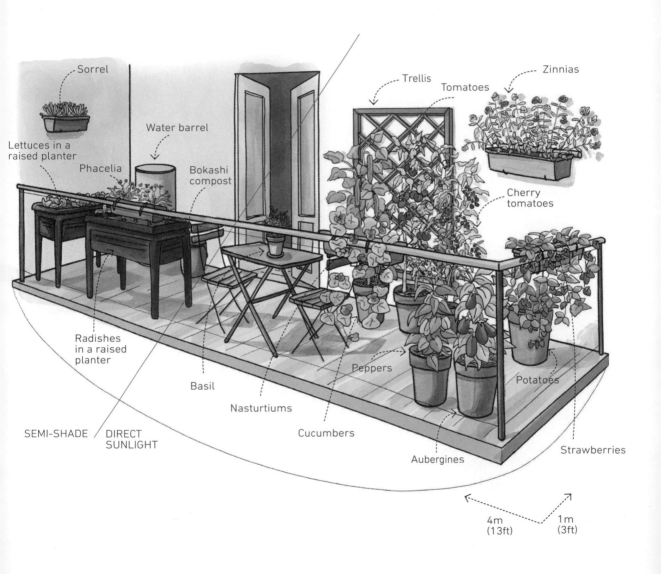

Sorrel

Water barrel

Lettuces in a
raised planter

Phacelia

Bokashi
compost

Trellis

Tomatoes

Zinnias

Cherry
tomatoes

Radishes
in a raised
planter

Basil

Nasturtiums

Peppers

Cucumbers

Potatoes

SEMI-SHADE / DIRECT
SUNLIGHT

Aubergines

Strawberries

4m
(13ft)

1m
(3ft)

FERTILIZING THE SOIL

Your plants are growing well, you've pollinated the first flowers, and you may even see the first fruits and vegetables starting to form. As the summer heat increases, you're going to see all these little beauties grow even faster, and this is when they will need more nutrients.

At the first signs of fruiting, it's important to feed your soil to provide more phosphorus for your tomatoes, peppers, aubergines, cucumbers, and strawberries. Phosphorus will not only nourish the plant and enable it to form new flowers, but let it know it's time to produce fruit or vegetables.

HOW TO ADD PHOSPHORUS

My favourite fertilizer is guano. This is bat and/or seabird droppings turned into manure in powder form. For an 18-litre pot (5 US gallons), take a shovelful of powder, pour it evenly over the soil, then water. Don't overdo it!

I use it most often at the start of fruiting, then add a second dose when I've started to harvest the first fruits and vegetables.

Guano may smell strong, but once it's diluted in the soil, it won't smell at all. Do cover your mouth and nose to avoid inhaling it!

Identifying phosphorus deficiency

Phosphorus deficiency can quickly affect your vegetable garden. On tomatoes, this can show up as purple spots on the leaves, and you'll notice a major slowdown in fruit development. So it's best to fertilize well and at the right time to avoid this.

A HIGH STAKES JOB

In the summer, your plants will start to grow faster, turning your balcony into a small urban jungle. That's where stakes come in!

WHAT'S IN IT FOR ME?

To make the most of your space, it's best to anticipate the strong growth of your plants and tie them up with sturdy vertical support.

Bear in mind that when a plant produces five or six tomatoes at once, it really will feel that extra weight. The risk of not using vertical support? A stem that gives way under the weight and breaks.

WHICH STAKES TO CHOOSE?

Use all the support you can find in your space. You can buy thick bamboo stakes, 1.5m (5ft) long. I usually fasten them with a cable tie to my balcony railings before I attach any plants to them. This provides a sturdy structure for the plants to carry their fruit or vegetables, but also prevents them from being toppled over by wind.

Saving time

There are metal structures you can attach to your pots to give your plants strong support. This is particularly effective for cucumbers.

PLANTS FOR FREE

Nature is generous! For instance, take your basil plant that grows so well: from a single plant, you can get two to four new ones, for free. It's a gift that keeps on giving.

BASIL PROPAGATION

Basil, like mint, can be propagated very easily – and infinitely. On page 75 you saw how to prune your basil plant in spring: each stem you cut can grow to a whole new plant.

1. Choose a healthy stem with several pairs of leaves, and cut it from the plant.

2. Keep a few leaves at the top of the stalk and remove the lower ones.

3. Take a shot glass or small coffee cup and fill it with water.

4. Put the basil stalk inside and place in a bright spot.

5. Change the water every two or three days.

6. In a week to 10 days, new roots will form. Plant the stem in a small pot of soil, and you're done!

LAYERING STRAWBERRY PLANTS

Once you've harvested the first fruits from your strawberry plants, they'll want to start reproducing. To do this, they will sprout long stems called runners, which will not bear fruit, but a new plant. These stalks are a bit like feelers coming out of your pots to try and make contact with fresh soil to start a new colony. It helps if you can lend them a hand:

1. Fill a small pot with potting soil and water it well to settle the soil and remove air pockets.

2. Place it next to one of your strawberry plants, then bend down the stem of a runner and gently push it into the pot with the leaves facing upwards and the young roots in the soil. Leave it attached to its parent plant.

3. Cover lightly with soil to keep the runner in place. Day after day, the young roots will grow. The runner acts as an umbilical cord, feeding the new plant.

4. After a week to 10 days, cut the runner near the base of your new strawberry plant. Once you have enough plants, cut off all remaining runners and any new ones that develop. This will "reprogram" the plant to resume fruit production. If you're lucky, you'll have a second strawberry harvest by the end of summer!

Good to know

If someone you know grows basil or mint, ask them for a small sprig. Just follow the technique described to grow it at home and produce a new plant. It also works really well with supermarket basil!

PRUNING IN SUMMER

While it's important to keep pollinating new flowers, you also need to continue pruning some of your plants regularly.

Aubergines In early summer, continue to remove side shoots.

Basil Continue pruning as explained on page 75. If you don't, the plant will grow flowers to set seed and reproduce, after which it will die back. Cut off any buds that do form to boost growth.

Cucumbers and peppers Cut off any leaves that turn yellow or look unhealthy.

Strawberries Cut runners once you have used what you need to make new plants.

Sorrel Harvest fresh new leaves as they grow. If they start to produce stems in order to flower, cut these off to encourage new leaves.

Tomatoes Continue to remove extra leaves at the base of the stems. Cut off any leaves that look diseased (they may turn yellow or purple, or simply dry out).

Potatoes There's nothing to do! Just wait patiently until the foliage and stems begin to wilt – this is a sign that it will soon be the right time to harvest your potatoes.

HARVESTING: THE KEY TO SUCCESS

Harvesting at the right time and on a regular basis is essential for healthy plants, but how do you know whether the fruit or vegetables are ripe?

WHY IS THIS SO IMPORTANT?

When you start a vegetable garden and see the first crops appear, you might get impatient and harvest too soon or, on the contrary, wait too long and harvest too late.

Harvesting at the right time will of course have a significant impact on the taste and texture of fruit and vegetables, but it will also enable your plants to have a longer, and more productive life. Each plant will redirect the energy and nutrients it was saving for the fruit you have just harvested towards those that are still growing.

SAVE THE DATE

All you need to do is learn more about the cultivars you're growing. A simple glance at your seed packet or a quick search on the internet will tell you the right time to harvest. That's why it's so important to label your seedlings right from the start!

Take a tomato, for example. When it turns red, you may think it's time to harvest it. This may be the case if the tomato is not too hard to the touch. However, some tomato cultivars will ripen while retaining a green, orange, and sometimes purple or black tinge, simply because of the characteristics of their specific cultivar. So it's very important to adapt harvesting to the cultivars of fruit or vegetable you're growing.

The case of the cucumber

For cucumbers, the average ripening size as shown on the seed packet will give a further indication of the harvesting period. There are cultivars producing large cucumbers, others that are long and thin, and others that will ripen while staying very small. In short, forget the stereotypical image of the long, plump, and completely green supermarket cucumber. The world of heirloom seed cultivars will open your experience to a whole new range of colours, flavours, and textures!

IDENTIFYING DISEASES AND DEFICIENCIES

Possible nutrient deficiencies or diseases are often visible through a change in leaf colour or texture. If identified early, they can be treated organically.

CURLED TOMATO LEAVES

This is a sign of overheating: the plant curls up to protect itself from the sun's rays.

Remedy Water regularly and provide shade during heatwaves.

PURPLE SPOTS ON TOMATO LEAVES

This is a sign of phosphorus deficiency (below, right). These spots often appear on the oldest, lowest leaves. You'll definitely notice them if the plant is producing a lot of tomatoes at the same time.

Remedy Add phosphorus to the soil by fertilizing with guano.

BROWN OR BLACK SPOTS ON TOMATO LEAVES

This can be a sign of tomato powdery mildew or downy mildew. These fungi develop on the leaves then gradually colonize the whole plant and kill it. They thrive in hot, humid weather.

Remedy Cut off infected leaves and discard them. Then clean your secateurs to avoid contaminating them with the fungus and spreading it to other plants. Spray the remaining leaves with a mix of 40 per cent cow's milk (a natural fungicide) and 60 per cent water. You can also buy Bordeaux mixture fungicide and spray it on the plant.

Curled tomato leaves

Purple spots on tomato leaves

BLACK SPOT ON THE BOTTOM OF THE TOMATO

A black spot on the underside of your tomatoes (below) is known as blossom end rot, and points to a lack of calcium and/or regular watering. The tomato will start to rot from the bottom upwards.
Remedy Calcium travels mainly through water, so it's important to water your tomato plants regularly. It's better to water a little every two days than to add large quantities of water every three to four days. Add calcium to your soil too.

YELLOW OR WHITE SPOTS ON CUCUMBER LEAVES

This is usually a sign of powdery mildew.
Remedy Same as for brown or black spots on tomato leaves.

YELLOWING LEAVES

You may see this on tomato, aubergine, or cucumber plants. If the leaf is yellow all over and without spots, this may indicate a general lack of nutrients.
Remedy Fertilize regularly with compost juice and add a little nitrogen.

ENTIRELY WHITE LEAVES

This is a sign of sunburn – our beloved plants don't wear sunscreen!
Remedy This happens to young plants that are not used to the sun. There's not much you can do apart from cutting off the sunburnt leaf.

Brown or black spots on tomato leaves

Black spot on the bottom of the tomato

Yellowing leaves

HUNTING DOWN SUMMER PESTS

Even in the summer, your plants can be attacked by certain pests – they too want a slice of the pie!

LEAF MINERS

If some leaves display white grooves, this is a sign they've been overrun by leaf miners. The tiny larvae of leaf miner flies make these marks by tunnelling through and eating the inside of the leaves.

Remedy Unfortunately, there's not much you can do apart from cutting off the affected leaves and discarding them.

SPIDER MITES

Are the leaves drying out? Have you found tiny spider webs? This is the work of spider mites. These small mites weave their webs between the leaves, feeding on their chlorophyll in the process. This will greatly hinder photosynthesis and cause the plant to wilt, as these colonies can develop very quickly. Spider mites might also attack other plants around them, so it's best to take this seriously from the outset.

Remedy Remedy Add 2 tbsp of organic soap and the juice of half a lemon mixed with water to a 1-litre sprayer. Spray the plant and leave for one hour, then rinse with water. Repeat the process until spider mites have gone.

Leaf miners

Spider mites

BIRDS

Are your tomatoes being nibbled on before you've even harvested them? Birds may have snacked on them. This may be no Hitchcock horror, but it does become quite a nuisance when it keeps on happening.

Remedy Put up a net around your tomato plants to protect them so birds won't be able to land near them. You can also hang objects that will move with the wind around your plants to scare the birds away. Old CDs will do the trick.

CATERPILLARS

Caterpillars can munch into your tomatoes and devour a salad in no time at all. You can also find their droppings on stems or leaves, which look like tiny black balls resembling fish eggs.

Remedy Add 2 tbsp of organic soap and the juice of half a lemon mixed with water to a 1-litre sprayer. Spray the plant and leave for for one hour, then rinse with water. Repeat the process until spider mites have gone.

RODENTS

They can sneak in anywhere and climb plants to devour your fruit and vegetables. They thrive in cities too, and won't turn down a little snack on your balcony!

Remedy A humane trap. Once captured, release the rodent far away from your home. Leave the trap for a few days to make sure other family members aren't still lurking around.

Birds   **Caterpillars**   **Rodent**

LOOKING AFTER YOUR FLOWERS

I'm sure your flowers have grown well and are a gorgeous addition to your balcony. I imagine they're also a delight for pollinators and other insects.

LOW MAINTENANCE NASTURTIUMS

Nasturtiums can't be pruned, so leave them alone. You can harvest a few flowers though – these pretty flowers are edible and will dress up any summer salad!

TWO-FOR-ONE ZINNIAS!

You can prune zinnias (below and right) to help them produce more flowers. Side shoots will grow, forming a V around the main stem. Cut the stem above the V to allow the shoots below to grow and produce more flowers. Use the flower you have removed in your own home arrangements!

NO-FUSS PHACELIAS

There's no maintenance required for phacelias. They're very popular with pollinators, who are attracted to the beautiful mauve flowers, and help them to set seed. When they die back, remove the dry stems, rake over the soil, and water. If you're lucky, you'll already have new seeds in your soil, which will start to grow over the course of the season.

WHAT TO PLANT IN SUMMER

KEEP AN EYE ON YOUR RADISHES

You can still plant radishes in the summer. Do harvest them early on though, because they will go to seed in the heat. When this happens, their colour will be slightly purplish, their texture stringier, and they're generally no longer good to eat. So keep an eye on them every day so they won't go to waste.

OPT FOR SUMMER LETTUCE

Plant more lettuce and switch to summer cultivars. Choose ones that will be used to hot weather and won't go to seed as quickly as spring lettuces.

HERBS

Plant your propagated basil sprouts: they'll love the sunlight and the heat! If you're enjoying the basil you might also like to start growing other kitchen herbs too, such as mint, thyme, chives, and sage.

SUMMER WATERING

Watering diligently is essential to see your vegetable garden through the summer – just one or two days' slacking off can cause serious damage to the most fragile plants. Better to be safe than sorry!

HOW DO I KNOW IF I NEED TO WATER?

Don't worry, your plants will soon let you know they're thirsty – their leaves will visibly start to droop. To avoid this, turn to page 67, where you'll find my instructions on how best to water your plants without wasting water.

Before watering, lift the mulch from your plant and check with your finger whether the soil in the pot is completely dry. If so, water the plant. If it's still cool and damp, wait until the next day.

APPROPRIATE WATERING

• **Tomatoes** Water sparingly but regularly to avoid blossom end rot in your tomatoes (see page 95). A little water every other day will do the trick.
• **Cucumbers** Water regularly and abundantly, as the plant is still growing and will need a lot of water.

WHAT TO DO IN THE EVENT OF A HEATWAVE

• **Do not water in the middle of the day** If you do, the water will quickly evaporate. What's more, cold water on warm soil could shock the plant. Wait until early morning or shortly before sunset to water.
• **Provide shade** Protect plants on your balcony from the blazing sun, moving them into the shade if possible.

• **Put saucers under your pots** This helps to retain water longer. Plants will stay hydrated by absorbing that extra water through their roots.
• **Do add mulch to keep the soil cool** Heat can cause it to decompose more quickly.

COLLECTING RAINWATER

Keep an eye out for summer rain so that you can collect rainwater and use less from the tap. Thanks to weather apps, you can easily find out when it's going to rain, and how much will fall.

If you don't have a water barrel connected to a gutter, improvise. Spread buckets, pots, saucers, and other containers on the ground. You'll be amazed at how much rainwater you can collect. Perfect for your plants, and even better for the environment!

Sustainable automatic watering

You can use a small water pump connected by drip hoses (which have tiny holes that drip water into the soil) to water each of your pots. Place the pump in a bucket of water, or better still in your water barrel, and set it up according to your plants' watering needs. This low-energy pump can also run on solar power.

By collecting rainwater or reusing shower water, I can easily provide enough water for the 18 pots and beds on my balcony without using tap water! Water is a valuable resource, and we can find simple and effective solutions to avoid wasting it.

WATERING YOUR VEGETABLE GARDEN WHEN YOU'RE AWAY

Unfortunately, I won't be available to plant sit while you're on holiday, as I have to look after my own plants! But here are some alternatives.

IRRIGATION POTS

These small terracotta containers can be buried in the soil around your raised garden planter. You can then fill them with water, and the porous terracotta will gradually drip the water down into the soil. The only drawback is their cost... And don't forget that these irrigation pots are only suitable for large containers such as garden beds. You can't put them in smaller containers.

THE WATER-BOTTLE TECHNIQUE

Plant upside-down water bottles in your pots, replacing the caps with ceramic cones (available from garden centres and online). These will slowly drip water into the soil. Terracotta cones are the best choice, as they have the same effect as irrigation pots. They're also less expensive.

AUTOMATIC DRIP IRRIGATION

There are automatic watering systems that you can program and connect to a water source. It's up to you to set watering frequency and amount. Test your system before you go to make sure you've set it up properly!

Drip irrigation is a tried-and-tested method for saving water and watering efficiently and evenly when you're away. You can also use the pump system described on page 101. However, if you're away for a long time and there's no more water in your water barrel or bucket, there will be no more watering. You can, however, ask a friend to pop over and fill it up.

Team up

Ask a neighbour, a relative, or a friend to come and water your vegetable garden – it's both free and convenient! Give them all the information they need before you leave and let them harvest and enjoy the ripe fruit and vegetables while you're away. You can offer to return the favour for them.

HARVESTING POTATOES

Are your potato plants starting to look yellow and wilted? Great, it's time to harvest them!

POTATO TREASURE HUNT

1. Grab a tarpaulin or plastic sheet and lay it on the floor. You can also reuse an empty bag of potting compost.

2. Water your potato plant thoroughly and wait for the water to drain.

3. Now turn your pot upside down onto the sheet, rifle through the soil and substrate, and reap the rewards of your labour!

4. Clean your harvest with water, then dry the potatoes thoroughly.

5. Store them in the dark for 2 weeks in a cool part of your home (such as a cellar), ideally at a temperature of 10–15°C (50–60°F) to preserve flavour and texture.

Turning green

Potatoes can develop solanine, a substance which gives them a green tint. Solanine protects them from pests or light, but it is toxic to humans. If you see even a slightly green spot, simply cut it out, discard it, and peel your potato – the rest is perfectly edible. On the other hand, if the potato has gone completely green, don't eat it.

HARVESTING SEEDS

If you've let a few lettuces or radishes go to seed, you can collect seeds to sow for next year's seedlings. You can do the same for tomatoes, aubergines, cucumbers, and peppers.

LETTUCE SEEDS

To state the obvious, a lettuce that goes to seed will produce ... seeds. In time you'll see a white down appear, a bit like a dandelion. That's where you'll find your seeds.

1. When your lettuce's foliage turns brown and the down is well formed, cut off the flower heads that are showing white down.

2. Place a sheet of paper on a table and shake each flower head over it. The seeds will fall out on their own.

3. Sieve through a colander to separate the seeds from the other plant material of the seed head, then slip them into an envelope.

4. Close the envelope tightly. Make sure to write down the name of the cultivar and the harvest date, then store your seeds in a dry, dark cupboard until you sow them in spring.

To obtain the best seeds...

Assuming you have used heirloom seeds, the harvested seed will carry the same genes as its parents. That's why you need to choose the healthiest, most beautiful fruits and vegetables to collect your seeds from.

TOMATO SEEDS

To get seeds with an impeccable pedigree, pick your best tomato, the star of your garden, with outstanding shape and colour. The best fruit are generally produced in the middle of the season, so avoid collecting seeds from the first and last tomatoes.

1. Select your tomato, and cut it in half.

2. Scoop out the pulp with a spoon.

3. Drop the seeds into a fine colander.

4. In your sink, run a gentle stream of water over the tomato seeds.

5. Each seed is surrounded by a protective jelly-like layer that needs to be removed.

6. To do this, place your seeds in a dish, add a little bit of water, and leave for two to three days. A layer of mould will appear, which is normal.

7. Pour the seeds into a colander.

8. Rinse: the protective coating will disappear.

9. Place the seeds on a paper towel and leave them to dry. Once dry, place the seeds in an envelope. Close tightly, note the name of the cultivar and the harvest date, then store in a dry cupboard, away from light until spring.

RADISH SEEDS

If you don't harvest the radishes, radish plants will develop long stems and form pods, in which the seeds are enclosed.

1. Let your plant wither until the pods turn brown: that's the time to harvest them.

2. Remove all the pods from their stalks and carefully open them.

3. Collect the seeds inside one by one, then put them in an envelope.

4. Seal the envelope tightly, note the cultivar name and harvest date, then store in a dark, dry cupboard until you are ready to sow them.

PEPPER, AUBERGINE, AND CUCUMBER SEEDS

1. Select a nice, ripe vegetable, then cut it in half and scrape the inside with a spoon to remove the seeds.

2. Place in a colander and rinse with water to remove the pulp from the seeds.

3. Place the seeds on a paper towel and leave them to dry.

4. Once dry, place the seeds in an envelope. Close it tightly, write down the name of the cultivar and the harvest date, then store in a dry, dark cupboard until the following spring.

WHAT ABOUT ANNUAL PLANTS?

By mid-September, plants that flower and fruit in one year will be reaching the end of their life cycle. Tomatoes, cucumbers, and aubergines will start to shed their leaves and become more vulnerable to diseases due to the cooler temperatures.

PARTING OF THE WAYS

Most of your summer plants will soon stop producing, and it's time to make way for autumn and winter cultivars. So, harvest the last of your fruit and vegetables and enjoy them before you enter a heartbreaking phase – you've shared your garden for months with these plants, they've amazed you, sometimes given you a hard time, but they've also borne wonderful fruits or vegetables. Now you're going to have to take them out of their pots and part with them.

Harvest green tomatoes

Are some tomatoes not quite ripe? Harvest them anyway! Put them all in a bowl with a banana and/or an apple and cover with a cloth. These fruits will release a natural gas, ethylene, which will ripen all your tomatoes in about 10 days.

Peppers in winter

Although cultivated as an annual plant in temperate climates, the pepper is actually a perennial that can live on for several years. So you could try placing your pepper plant by a bright, warm window and letting it spend the winter indoors. Water it when the soil gets dry, and don't be surprised if the leaves fall off. When spring comes, the plant will grow them back and you'll have peppers much earlier than usual! Watch out for aphids, though – your pepper plant will be very tempting indeed.

REMOVING ANNUALS FROM POTS

• Cut the main stem about 10cm (4in) from the soil and leave the stump in the container for a good week to allow the roots to start decomposing. This will make it much easier to remove the roots from the pot.
• Cut healthy foliage and stems into small pieces, then add them to your compost.
• For tomatoes, you can also recover healthy foliage for use as fertilizer (see page 74).

PREPARING FOR AUTUMN

At the end of summer, it's important to get ready for autumn. You will be starting seedlings for the new season: so long tomatoes, hello spinach!

THE AUTUMN AND WINTER VEGETABLE GARDEN

• **7 pots** Pak choi, kale, fennel, chard, parsley or coriander, cauliflower, broccoli.
• **5 window boxes** Strawberry, spinach, sorrel, phacelia, lamb's lettuce.
• **2 raised planters** Winter salad and mizuna, turnips. Sorrel and strawberries are perennial plants, so they can stay in their pots, just like phacelias. If you protect them, they'll grow back next spring!

STARTING AUTUMN SEEDLINGS

Unlike spring and summer seedlings, which are sprouted indoors, autumn and winter seedlings can be sown outdoors, as temperatures are just right for proper germination and growth.

Start your seedlings as early as the second or third week of August. They will be ready for planting two to three weeks later.

As you did in spring, feel free to sow more seeds than necessary, to make sure you have enough seedlings to plant.

My favourite seeds

• **Broccoli** 'Purple Peacock' or 'Waltham'
• **Cauliflower** 'Verde di Macerata' or 'Neckarperle'
• **Chard** colour mix
• **Fennel** 'White Perfection'
• **Flat-leaved parsley**
• **Kale** 'Halbhoher Grüner Krauser'
• **Lamb's lettuce** 'Verte à cœur plein'
• **Mizuna** or **Japanese mustard**
• **Pak choi** 'Xianggang heiyebaicai' or 'Tatsoi Spoon'
• **Spinach** 'Géant d'hiver'
• **Turnip** 'Globe à collet violet'
• **Winter lettuce** 'Rougette de Montpellier'

HOW TO SOW AUTUMN SEEDLINGS

All sowing can be done in the regular way, with two exceptions:

• **Chard** When germinating, two or three stems will grow from a single seed. After allowing them to grow for a week, keep the strongest stem, and remove the others by gently pulling on them.
• **Lamb's lettuce** You can sow these directly by scattering seed in a trough or window box around mid-September.

A few more tips

• Avoid placing your seedlings in direct sunlight, especially on hot days.
• Don't forget to keep the soil moist to help your seedlings sprout.
• Place seedlings in a place that gets enough sunlight so they don't grow too long.
• Avoid placing your seedlings directly on the ground: set them on a table or windowsill instead to keep them away from slugs and snails.

CROP ROTATION

Now's the time to rotate your summer varieties with your new seedlings. This table shows you how to go about it and the quantities of seedlings required.

	Spring–summer plants	Autumn–winter plants	Volume
Pots	Basil	Replace with 1 parsley or coriander seedling	1 pot of 5 litres (1.5 US gallons)
	Cherry tomato	Replace with 1 fennel seedling	1 pot of 12 litres (3 US gallons)
	Aubergine, cucumber, pepper, potato, tomato	Replace with 1 broccoli, chard, kale, and cauliflower seedling, and 3 pak choi seedlings	5 × 18-litre pots (5 US gallons)
Window boxes or troughs	Strawberry	Leave in place	1 × 15-litre planter (4 US gallons)
	Sorrel	Leave in place	1 × 15-litre planter (4 US gallons)
	Phacelia	Leave in place	1 × 15-litre planter (4 US gallons)
	Nasturtium	Replace with broadcast lamb's lettuce seeds	1 × 15-litre planter (4 US gallons)
	Zinnia	Replace with 8 spinach seedlings	1 × 15-litre planter (4 US gallons)
Raised planters	Lettuce	Replace with 4 winter salad seedlings and 6 mizuna seedlings	1 × 50 or 100-litre raised planter (13 or 26 US gallons), depending on available space
	Radish	Replace with 8 turnip seedlings	1 × 50 or 100-litre raised planter (13 or 26 US gallons), depending on available space

PREPARING POTS

Rather than buying new potting soil, try to reuse as much summer potting soil as possible.

REUSE SUMMER POTTING SOIL

After a week or so, the roots of the plants you cut should have started to decompose. Now it's time to remove them and prepare your pots for new plants.

1. Pull out the remaining piece of the stem, which will come with a few roots. Shake it over the pot, so that the root-free soil falls back in.

2. Take a small rake or fork, and rake across the pot to remove any bits of root that may remain.

3. Add coffee grounds to the soil at a ratio of about 1 to 20, and mix with the potting soil.

4. Spread around a little new potting compost to make up for the lost volume.

5. Add 2 trowels of material from your compost, bury deeply, and cover with potting soil.

6. Level, water, and leave to stand for a few days before planting your autumn seedlings.

PLANTING AUTUMN SEEDLINGS

Ideally, plant your seedlings in mid-September, when temperatures are still warm enough for them to start growing well. The quantities to be planted per pot and raised garden bed are listed in the "Crop rotation" table on the opposite page.

AUTUMN

AUTUMN ON YOUR BALCONY

Autumn is here! A new cycle begins in the vegetable garden, with beautiful seasonal cultivars that will bring you lots of joy and bountiful harvests.

EMBRACING CHANGE

The landscape changes, and the trees' foliage turns from green to brown, red, yellow, or orange. As the first leaves begin to fall, collect them and use them in the vegetable garden (see page 118).

Plant growth will gradually slow down. Temperatures become cooler, and the sun's heat is milder, but autumn cultivars will adapt very well. They need cooler weather to thrive.

SHOPPING AND REPURPOSING LIST

- Protective fleece
- Gathered dead leaves
- Old lids, reused plastic bottles

AUTUMN TO-DO LIST

- Organize the balcony space for autumn
- Fertilize
- Continue composting
- Apply mulch
- Protect plants from pests
- Harvest the first autumn crops
- Anticipate the first frosts, and protect the vegetable garden

KEEP AT IT!

Although watering will become less of a chore with more rainfall, your vegetable garden will keep you busy in a different way. You'll especially need to be on the lookout for slugs and snails, which can proliferate at this time of year. The first frosts will also start to arrive. You'll need to anticipate all this to protect your crops!

ORGANIZING YOUR BALCONY FOR AUTUMN

When autumn arrives, most of your vegetables will need sunshine to thrive, and to enable them to cope with the cooler temperatures.

MOVERS AND SHAKERS

To help your plants make the most of autumn sunshine, it's best to move your raised planters and most of your pots to the sunniest spot on the balcony.

You can leave your strawberry planter in the sun. The leaves will turn brown, but don't worry – your strawberry plant will still be alive and well in the spring after a good pruning. For the time being, let it go into hibernation.

SEMI-SHADE

Some plants will need to stay in the semi-shade to thrive, such as cauliflower, sorrel, lettuces, and mizuna. Phacelia also needs semi-shade; it will set seed and die back, growing again from seed next year without you having to do anything.

If you still have room, feel free to add more pots if you have some seedlings left!

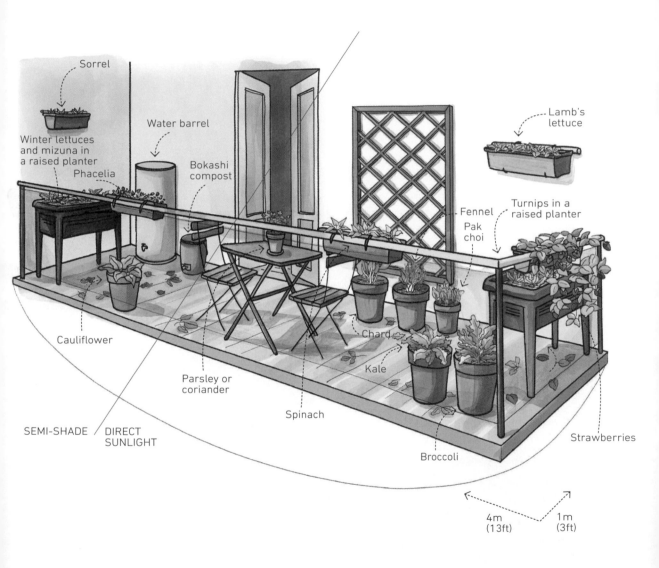

Sorrel

Winter lettuces
and mizuna in
a raised planter

Phacelia

Water barrel

Bokashi
compost

Lamb's
lettuce

Turnips in a
raised planter

Fennel

Pak
choi

Cauliflower

Parsley or
coriander

Chard

Kale

Spinach

Strawberries

Broccoli

SEMI-SHADE | DIRECT
SUNLIGHT

4m
(13ft)

1m
(3ft)

MULCH AND FERTILIZE

In the summer, mulch is mainly used to improve water retention in the soil and prevent weeds from growing. In autumn, it acts as insulation against the cold – essential to help your plants face the first frosts!

GATHERING DEAD LEAVES

Good news! In autumn, you can easily find healthy, natural, free, and readily available mulch – dead leaves. All you have to do is fill a large bag with any leaves (avoid horse chestnut, oak, and black walnut leaves) that have fallen on pavements and other hard surfaces. If left where they are, these leaves will decompose without nourishing the soil.

Leaves that have landed on grass or under a tree, on the other hand, will fertilize the plants they have landed on, so these should be left where they are.

Once you've finished collecting the leaves, crush them in your hands and add them as a layer in your pots and planters. Of course, you can always use straw or grass clippings as a mulch if those are easier to obtain.

Saving money

You can also use your own hair for effective, long-lasting mulching. It contains a large amount of nitrogen and takes a considerable time to decompose, thanks to its keratin content. There is one thing to bear in mind, however – use only natural hair that has not been dyed or had chemicals applied to it. When you go to the hairdresser, feel free to ask to pick up your cut-off hair – most of the time, it will be thrown away anyway.

You can also use the fur left on the brush after grooming your cat or dog.

FERTILIZING IS HEALTHY!

Two or three weeks after planting your autumn seedlings, the application of a little fertilizer will go a long way.

• **Compost juice** Feel free to give your plants nitrogen and potassium to ensure good rooting, beautiful leaf growth, and a generally healthy structure.
• **Nettle manure** This can be used to boost your plants' immune systems. You can also add natural seaweed fertilizer or reuse your own urine (see page 70).
• **Dead leaves** If you bury them deep within the compost before planting your seedlings, these will enrich your soil.

AUTUMN PESTS

Every season has its pests, although some, like snails and slugs, can be found all year round. Autumn is a season in which they proliferate, as cooler, wetter weather provides ideal conditions for these invertebrates, which are fond of small, green shoots.

PROTECT AGAINST SLUGS AND SNAILS

Cover your seedlings with plastic bottles or clear food containers, or better still, with purpose-made lids that fit your pots and garden beds. Ensure there is ventilation, making holes in the covering if necessary.

Sometimes snails and slugs manage to get past the barrier, or there may already be eggs in your potting soil. So inspect your plants every evening, and remove any that you find.

A beer trap is a great way to prevent slugs and snails from proliferating. Simply add some beer to a small container placed on the soil, and the slugs will be attracted and fall in. The idea is not to wipe them out, but to keep your vegetable garden from being completely overrun by a large and hungry snail and slug population.

KEEPING AN EYE ON APHIDS

Even when temperatures get cooler, aphids may still be around. Inspect your plants carefully and treat as explained on page 72. Pouring rainwater or tap water onto the aphids, or using a spray of soapy water will get rid of them.

CABBAGE ROOT FLIES

Cabbage root flies love to attack cauliflowers and turnips and can really be a nuisance. Much like the common housefly, it lays eggs in vegetables which then hatch into tiny larvae. To feed, the larvae nibble the roots of the host plant, digging tunnels and damaging the plant.

Because it's better to be safe than sorry, prevent cabbage root flies from laying eggs in your pots by covering turnips and cauliflower with an insect-proof mesh.

AUTUMN HARVESTS

The first autumn harvests are just around the corner. You can enjoy winter lettuce, mizuna, pak choi, spinach, and kale.

GOOD THINGS COME TO THOSE WHO WAIT

You'll have to be patient while waiting to harvest your fennel, chard, and turnips as they need more time to grow. Some may be ready before the onset of winter, while others will quietly make it through the cold season and be ready to harvest the following spring. Keep a close eye on them and protect them from pests.

LEAFY VEGETABLES

It's best to harvest leaf by leaf, to allow the plant to keep producing. Spinach, for example, will grow new leaves regularly, as will mizuna, and pak choi.

If you'd rather harvest the whole plant, that's also an option. You can plant a new seedling immediately after harvesting the whole vegetable (see page 108). However, bear in mind that the later you plant, the slower it will grow. In fact, cooler temperatures will often encourage the plant to become dormant. It will then save as much energy as possible to survive the cold rather than using it to grow.

PROTECTING PLANTS FROM FROST

You can effectively protect your plants from frost if you plan ahead and follow a few simple steps.

BETTER SAFE THAN SORRY!

Keep an eye out when the first winter frost date gets closer. Check the weather forecast so that you can prepare for frost at night and protect your plants.

Protective fleece is one readily available option. These white pieces of fabric can be draped or tucked over your plants, letting sunlight through while preventing frost from damaging the leaves.

THE RIGHT STEPS AND THE RIGHT TOOLS

• **Choosing the right location** On a balcony, there are a number of strategic locations where plants can best be protected from the frost. Ideally, they should be placed close to a wall that can provide warmth and protect them from cold winds.

Do you need to protect all your plants?

Some plants are actually incredibly resilient to cold and frost. For example, parsley and spinach can withstand temperatures as low as -15°C (5°F), while kale can withstand temperatures as low as -20°C (-4°F). Impressive!

• **Plants resistant to a light frost, around -2°C (28°F)** Pak choi, cauliflower, fennel, turnip, chard.

• **Plants resistant to heavier frost, below -2°C (28°F)** Kale, parsley, spinach, lamb's lettuce, winter lettuce, mizuna.

- **Avoid watering, and empty the saucers** If the soil is wet and frost sets in, this can freeze the potting soil and the entire pot.
- **Raising pots** Elevating your pots on a brick, for example, will prevent the plant from coming into contact with the ground when it gets too cold.
- **Protective fleece** It's inexpensive, effective, and can be reused every year. Make sure each plant is properly covered, then secure your fleece by placing your pot on the loose ends.
- **Covers** These plastic lids with adjustable ventilation are a great solution to protect your plants against the frost and are available for both pots and raised garden beds.

Should I bring my plants indoors?

In the event of severe frosts, you may choose to bring a few plants indoors temporarily. However, be careful not to expose your plants to very warm temperatures if they have already been subject to frost.

EARLY
WINTER

EARLY WINTER ON YOUR BALCONY

Winter brings cold and frost in your vegetable garden. However, it's still an important season for nature, which will benefit from higher rainfall to replenish the water table. If you've got a rainwater barrel, now's your chance to fill it up!

A SLOW-MOTION VEGETABLE GARDEN

Let's face it, winter isn't the most productive of seasons in the vegetable garden. Everything seems to come to a standstill, as if nature had pressed the pause button. Still, even though your harvests may be small, this will allow you to appreciate your seasonal produce even more. You may feel nostalgic about your beautiful green summer jungle, but don't worry, it'll be back in a few months' time!

SHOPPING AND REPURPOSING LIST

• A seed tray to grow your own sprouted seeds, or reuse a glass jar

• A microgreen growing kit, or reuse plastic containers

EARLY WINTER TO-DO LIST

• Continue to protect plants from the cold

• Enjoy winter harvests

• Care for indoor plants

• Try microgreens

• Grow your own sprouts

• Clean pots and protect gardening equipment

• Sort and store seeds

• Take notes on what this year's gardening has taught you

• Prepare for the next seasons and the seedlings to be started soon

EAT SEASONALLY

This is all the more important in the winter, when winter vegetables provide the nutrients our bodies need. Our bodies no longer crave fruit with a high water content as they do in the summer, so you can stop buying tomatoes! Enjoy broths, soups, and delicious baked vegetables. You can even experiment with indoor gardening by growing microgreens and sprouted seeds, which are both delicious and very nutritious.

ORGANIZING YOUR BALCONY FOR WINTER

The vegetable garden survives the cold and frost by going dormant. Now is the time to take a few simple precautions to help your plants get through the season.

STAYING WARM

• **Be prepared** Place as many plants as possible against your walls to shelter them from the wind and cold. Identify areas of your balcony that are more exposed than others and avoid placing your plants there. You don't need to move parsley, spinach, and kale, which are all highly resistant to frost.

• **Frost** Don't take any chances when frost is forecast, and use lids to cover your garden beds and protective fleece for your pots. This won't damage your plants but will provide them with the protection they need.

• **Sunny days** Place all your pots in the sun to let them enjoy some of the light and warmth.

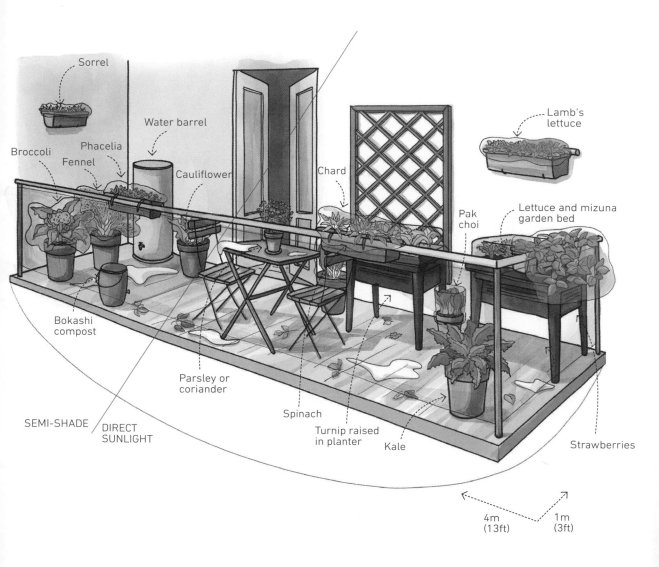

Sorrel

Water barrel

Broccoli

Phacelia
Fennel

Cauliflower

Chard

Lamb's
lettuce

Pak
choi

Lettuce and mizuna
garden bed

Bokashi
compost

Parsley or
coriander

Spinach

Turnip raised
in planter

Kale

Strawberries

SEMI-SHADE

DIRECT
SUNLIGHT

4m
(13ft)

1m
(3ft)

WINTER HARVESTS

In the winter, everything grows much more slowly, but you can still get small harvests, mostly from leafy vegetables.

SUCCULENT TURNIPS

Root vegetables like turnips are ready for harvest. Depending on the cultivar, you'll find the right size and colour for ripeness on the seed packet. Harvesting couldn't be easier: just pull them out by grabbing the base of their stems, like a large radish! Enjoy raw and thinly sliced, with olive oil and sesame seeds, or baked in the oven.

Spa day

While frost will naturally soften kale leaves and improve their texture, you can also massage them. Bond with your kale plant, and its leaves will be even more tender.

HOW'S YOUR KALE?

If you're lucky enough to be able to harvest some kale leaves, the texture will be even better after the first frosts, as this will tenderize them. When it comes to cooking kale, you'll have plenty of options. I love to eat kale raw in salads, stir-fried, or as crisps.

CHECKING UP ON CHARD

Your chard leaves will also have reached maturity. They are delicious in salads, but can also be stir-fried, much like spinach.

BROCCOLI, CAULIFLOWER, AND FENNEL

You'll have to wait a little longer, but when the harvest is ready, you're in for a treat. All can be picked when they are a good size and firm to the touch. The foliage of these vegetables is also edible. Fennel sprigs can be used as a herb. In addition to the cauliflower head, you can use the leaves to make a stir-fry in a wok or pan, or a hearty soup. Lastly, every part of broccoli can be eaten, even the stem!

Don't throw away turnip greens!

The turnip's fresh, dark green foliage is perfectly edible, and there are simple recipes for turning it into a delicious homemade pesto or soup. The perfect winter dish!

CARING FOR PLANTS INDOORS

If you've taken some of your plants or fruit trees indoors for the winter, be sure to give them a prime spot near a sunny window.

INDOOR PLANT MAINTENANCE

There's no need to feed your plants in winter, but water them regularly so that the soil in the pots doesn't dry out completely. Place saucers under the pots to retain excess water (though don't water them too much) and prevent your living room from turning into a swimming pool.

The foliage may dry out and many leaves may fall. Don't panic: it's a perfectly normal cycle. When the sun comes out, a healthy plant will grow brand-new foliage.

KEEP YOUR EYES PEELED FOR PESTS

• **Woolly aphids** These form small white clusters resembling wool or mildew. To remove them, brush affected branches and leaves with a toothbrush and organic soap.
• **Aphids** They can spend winter on your plants and cause the same damage that they would outdoors. If you spot them, treat immediately (see page 72).
• **Spider mites** They can also be a problem indoors, so keep an eye out for small spider webs and treat them promptly if you see them (see page 96 for treatment).

My pots are overrun with midges!

Sciarid flies are tiny midges that lay their larvae in the soil of indoor plants. To treat them, spray with a mix of water and Castile soap. You can also buy nematodes, which are natural parasites that will eat midge larvae in your soil. Simply mix the nematodes with water and water your soil with this solution.

TRY MICROGREENS

Even if outdoor gardening is on hold, you can still garden indoors by trying your hand at growing microgreens — young seedlings of vegetables and other plants that are full of nutrients.

TINY BUT MIGHTY

You can grow microgreens using a wide variety of seeds, from radish, lettuce, broccoli, and peas to sunflower, rocket, and alfalfa.

They should be harvested as young seedlings, as they contain an enormous amount of nutrients, sometimes more than a mature vegetable. Ideal for recharging your batteries in winter!

HOW TO GROW MICROGREENS

You can buy ready-to-use kits for growing microgreens, but it's also possible to repurpose plastic packaging and use it as a tray for growing your own seeds.

1. Upcycle plastic containers such as sushi trays or other leftover plastic containers. Poke small holes in the bottom so that the water can drain.

2. Spread a thin 5cm (2in) layer of potting compost over the base of the container. You can also mix potting compost with coconut fibre, which will greatly improve water retention.

3. Water the potting compost lightly.

4. Scatter the seeds. Use a generous quantity of seeds as one seed will produce one microgreen.

5. Firm the soil using a lid or other flat tool so the seeds lie on the potting mix. Water with a spray bottle so that you don't dislodge the seeds.

6. Put a lid on the container, then place the microgreens near a bright window or under a grow lamp if you have one. You can also leave them in complete darkness for the first three days, as this will force them to form stronger roots.

7. Water daily with the sprayer to keep the soil moist.

8. After a few days, the first shoots will appear. You can now remove the cover.

9. When they reach a height of 10cm (4in), it's time to harvest them. Snip them off with scissors.

How do you eat them?

Microgreens are a great pairing with salads, fish, or homemade sandwiches. Personally, I love a slice of toast with salted butter and a handful of microgreens. It's up to you to experiment and give them a try!

HOMEGROWN SPROUTED SEEDS

Even easier than microgreens, sprouted seeds are also well worth a try. You can experiment with the same seed varieties as microgreens, but without using potting compost, as sprouted seeds are simply grown in water.

WHAT DO I NEED?

All you need is a germinator, or make one from a glass jar with a wire-mesh or cloth lid, and a container to catch the water as it drains.

GIVE IT A TRY!

1. Pour a thick layer of seeds into a jar until the bottom is completely covered.

2. Add water and leave overnight.

3. The next day, cover with a wire-mesh lid or a piece of cloth held in place with a rubber band, and turn the jar upside down to drain the water.

4. Then place the jar upside down and slightly tilted on a container (such as a bowl or tray) that will catch the water as it drips out. In the evening, run the jar under the tap to rinse your seeds, and drain the water. Return it to its stand, upside down and tilted. Repeat this operation twice a day to rinse your seeds.

Saving Time

Health food shops stock germinators, which are both inexpensive and practical. Glass germinators are preferable.

Saving Money

Reuse a glass jar with a thin piece of cloth or gauze, which you can secure to the jar with a strong rubber band.

5. In a few days, the seeds will start to sprout.

6. When your jar is full, you can harvest the sprouted seeds. In just ten days, you'll have a jar full of them to enjoy with salads, fish, or sandwiches. A nutritious treat!

HOW LONG DO THEY KEEP?

You can store sprouted seeds for several days in an airtight container in the fridge. Just after harvesting, I recommend draining them and placing them on a paper towel for a few minutes to dry before storing.

Homegrown sprouted seeds **135**

TAKING CARE OF YOUR EQUIPMENT

It's chilly outside, and I imagine you're not spending much time on your balcony. Nevertheless, keep protecting your plants, and take time to spruce up your equipment.

CLEAN UNUSED POTS

If you've harvested everything outdoors and have pots standing empty, you can clean them by scrubbing with organic soap and water. It's vital to do this before the next season to remove fungus residue or pest eggs that could later contaminate the soil or damage your delicate young plants.

EQUIPMENT MAINTENANCE

Garden equipment maintenance is essential, especially to protect your secateurs or knife from rust. Avoid leaving your tools outside exposed to rain or frost. If they do get rusty, use a rust remover to clean them.

Use a whetstone to carefully sharpen your tools – there's little more annoying than using dull secateurs and knives.

SORTING AND STORING SEED

Being organized will save you a lot of time next season.

ORGANIZE YOUR SEEDS BY SEASON

Look at the recommended sowing times and organize your seeds accordingly. This means you'll have all the seeds you need for each season at your fingertips.

MAKE AN INVENTORY

As you sort them, you can also keep track of how many seeds you have left per plant variety. This will let you know whether or not you need to buy new seeds.

You'll also be able to draw up a calendar for harvesting your seeds over the coming seasons. For example, if your stock of lettuce seeds is running low, you can simply let a lettuce go to seed next season to replenish your stock of lettuce seeds.

TRY NEW VARIETIES

You probably have your favourites by now, but perhaps you'd still like to try other cultivars next year? For example, if you enjoyed growing aubergines this year, why not try several cultivars next year? There are so many interesting and different vegetable cultivars offering various colours, shapes, sizes, and flavours. You can start looking for inspiration on heirloom seed sellers' websites. It's also a great time to organize a seed swap with gardener friends!

Take stock of the past year

As you sort through your various seeds, assess the past year by asking yourself: "So would I grow this again?" The answer will depend on your preferences but also on your experience growing this variety. Perhaps it wasn't productive enough, or perhaps its texture or taste wasn't to your liking.

TAKING STOCK FOR NEXT YEAR

It's time to take stock of your first year of vegetable gardening on your balcony. Your experience is unique, because you've learned to adapt to your cultivation conditions (climate, space, local biodiversity...). Take the time to sit down and make notes about all your observations.

• **What was your biggest mistake?** It's important to analyze why, so you can learn from it and avoid repeating it.

. . . .
. . . .
. . . .
. . . .

• **If you had to do it all over again, what would you have done differently?** This might be organizing your balcony space, tending to your compost, planting calendars, and so on.

. . . .
. . . .
. . . .
. . . .

A FEW QUESTIONS TO GUIDE YOU

Answering these questions will help you to plan your balcony garden for the coming seasons. The most important thing is to try and understand where you failed and where you succeeded – that's how you make progress!

• **What has been your greatest success?** Why?

. . . .
. . . .
. . . .
. . . .

• **Can you optimize your space even further to expand your vegetable garden?**

. . . .
. . . .
. . . .
. . . .

• **Can you use your resources (water, potting soil, fertilizers, etc.) in a more efficient, less wasteful way?**

. . . .
. . . .

• **Fertilizers** You won't be able to make your own guano. But for other fertilizers, feel free to make them yourself by utilizing the resources available to you, which can be just as effective.

• **Compost** In addition to bokashi compost, you can also try a worm bin using earthworms. These two methods are highly complementary: worm bins will give you new potting soil, which you can enrich with the juice from the bokashi compost.

Do your homework – but do it with joy, and feel free to cheat and share your answers with other gardeners!

EXPERIMENT WITH NEW FRUITS AND VEGETABLES

Why not start with courgettes in pots? An 18–20 litre (5–5½ US gallon) pot will work just fine.

As well as strawberries, maybe raspberries would be nice too?

And why not experiment with new flowers, or even try to grow fruit trees? There are dwarf fruit trees that are perfectly suited to growing in pots and can be very productive. This is particularly true of the calamondin orange tree, which I highly recommend!

SAVING RESOURCES

Natural resources are becoming scarcer, so it's up to us to lessen our impact in our everyday lives as well as in the vegetable garden by asking ourselves the right questions.

• **Water consumption** Are there ways you can reuse more water?

• **Potting soil** Buy it only when you need it. You can reuse and enrich potting soil over several seasons before ordering a new bag.

INDEX

BIBLIOGRAPHY

4 Foreword
Intergovernmental Panel on Climate Change (IPCC), (2022). "AR6 Climate Change 2022: Mitigation of Climate Change", https://www.ipcc.ch/report/sixth-assessment-report-working-group-3
11 Connecting with food and the seasons
Brian Halweil, (2007)."Still No Free Lunch: Nutrient levels in US food supply eroded by pursuit of high yields", Organic Center

ACKNOWLEDGMENTS

Thank you for holding this book in your hands and for reading it. Welcoming nature back into our lives and our cities is an essential part of protecting our environment, our generation, and, above all, future generations. Starting a vegetable garden is a form of activism, helping us reconnect to both food and the seasons. Thank you all!

This book is for all the beautiful souls and dreamers who aspire to a fairer world and a healthier planet. Let's keep dreaming, shaking up our habits, asking questions, and uniting: every little gesture counts.

I'd like to thank my Junis for her love, generosity, kindness, and unfailing support. I am honoured and blessed to share this life with you. Thank you for encouraging me to follow my dreams. And Amaury, I'm already proud of you.

Many thanks to my family, my parents Dominique and Daniel, my sister Aude, my aunt Giovanna, and my cousin Marie-Anne, for their regular encouragement, their love and support as I turned my life and aspirations around.

I would also like to thank another Dominique, who will recognize himself. Thank you for sharing your love and respect for nature with me.

I am extremely grateful to Eduardo Terzidis and Sarah Wu, my permaculture teachers and friends, who taught me a passion for this wonderful life philosophy.

An emotional thought for Cosmo. Thank you for your boundless love and the life lessons you taught me. You will always be in my heart.

Many thanks to all my close friends who have been with me for so many years: Marie-France, Charlotte, Pierre, Flo, Charouf, Filou, Bibouille, Maxou, Laurita, Francouze, Fouad, Cécile, Yann, Sebbi, Blazou, to name but a few.

And thank you, Julia, the amazing photographer who bent over backwards to capture a tomato or cucumber at its best angle!

Finally, I'd like to thank Larousse, and especially Sylvie, Philippine, Nathalie, and Grégoire, for believing in me and giving me the chance to write this book.

Sharing my passion for vegetable gardening with you has been a great pleasure, and I hope it's been infectious! Now it's up to you to pass on the torch to others, and inspire younger generations too.

If you enjoyed this book, please feel free to leave a review online to support my work, and send me a message on Instagram **@TheFrenchieGardener** or on my website **www.thefrenchiegardener.com**

Green love!

ABOUT THE AUTHOR

Patrick Vernuccio, The Frenchie Gardener, is a French author, facilitator, and content creator on urban balcony gardening. Passionate about permaculture, he strives to adapt his philosophy and techniques to urban environments, to once again welcome the taste of nature in our lives and in our cities! Patrick has more than one million followers on Instagram, TikTok, and YouTube: @thefrenchiegardener

PUBLISHER'S ACKNOWLEDGEMENTS

Dorling Kindersley UK would like to thank Jane Simmonds for editing this book, John Friend for proofreading, and Hilary Bird for indexing.

PICTURE CREDITS

© **Patrick Vernuccio** All images
© **Julia Castel** Pages 3, 5, 7, 10, 16, 19, 33, 44, 46, 59, 62(l), 64, 80, 81(t), 83, 86, 89, 92, 93, 99, 112, 116, 118, 123, 143
© **Liz Eve** Pages 8, 9, 11, 14, 15, 82, 83, 101
© **Lucas Blazek** Page 97(l)
© **Shutterstock** Pages 12, 35, 38(t), 43(b), 52(t), 69, 72, 78(b), 95(l), 96, 97, 120(r), 131, 136, 137, 139(t)

LAROUSSE
Publishing Directors
Isabelle Jeuge-Maynart, Ghislaine Stora
Editorial Directors Émilie Franc, Julie Martin
Artistic Director
Géraldine Lamy
Publishing Managers
Sylvie Cattaneo-Naves, Philippine Richard
Editor Sophie Jutier
Cover Designer Valentine Antenni
Graphic Design, Illustrations, and Layout
Claire Morel Fatio
Production Émilie Latour

DK LONDON
Editorial Manager Ruth O'Rourke
Assistant Editor Jasmin Lennie
Production Editor Tony Phipps
Production Controller Stephanie McConnell
Art Director Maxine Pedliham
Publishing Director Katie Cowan

DK INDIA
Managing Art Editor Neha Ahuja Chowdhry
DTP Designers Manish Upreti and Umesh Rawat
DTP Coordinator Pushpak Tyagi
Pre-production Manager Balwant Singh
Creative Head Malavika Talukder

This edition published in 2024 by
Dorling Kindersley Limited
DK, One Embassy Gardens, 8 Viaduct Gardens,
London, SW11 7BW

First published in 2023 by Larousse
21 rue du Montparnasse, 75006 Paris, France

The authorised representative in the EEA is
Dorling Kindersley Verlag GmbH. Arnulfstr. 124,
80636 Munich, Germany

A CIP catalogue record for this book
is available from the British Library.

ISBN: 978-0-2416-7774-2

Printed and bound in China

www.dk.com

MIX
Paper | Supporting
responsible forestry
FSC™ C018179

This book was made with Forest Stewardship Council™ certified paper – one small step in DK's commitment to a sustainable future. Learn more at
www.dk.com/uk/information/sustainability